SOIL CONSERVATION

ASSESSING THE NATIONAL RESOURCES INVENTORY

Volume 2

Committee on Conservation
Needs and Opportunities

Board on Agriculture

National Research Council

NATIONAL ACADEMY PRESS
Washington, D.C. 1986

National Academy Press • 2101 Constitution Avenue, NW • Washington, DC 20418

NOTICE: The project that is the subject of this report was approved by the Governing Board of the National Research Council, whose members are drawn from the councils of the National Academy of Sciences, the National Academy of Engineering, and the Institute of Medicine. The members of the committee responsible for the report were chosen for their special competences and with regard for appropriate balance.

This report has been reviewed by a group other than the authors according to procedures approved by a Report Review Committee consisting of members of the National Academy of Sciences, the National Academy of Engineering, and the Institute of Medicine.

The National Research Council was established by the National Academy of Sciences in 1916 to associate the broad community of science and technology with the Academy's purposes of furthering knowledge and of advising the federal government. The Council operates in accordance with general policies determined by the Academy under the authority of its congressional charter of 1863, which establishes the Academy as a private, nonprofit, self-governing membership corporation. The Council has become the principal operating agency of both the National Academy of Sciences and the National Academy of Engineering in the conduct of their services to the government, the public, and the scientific and engineering communities. It is administered jointly by both Academies and the Institute of Medicine. The National Academy of Engineering and the Institute of Medicine were established in 1964 and 1970, respectively, under the charter of the National Academy of Sciences.

This project was supported under Agreements No. 59-3A75-4-57, Soil Conservation Service, and No. 59-32U4-4045, Science and Education, between the U.S. Department of Agriculture and the National Academy of Sciences.

Library of Congress Catalog Card Number 86-60330

ISBN 0-309-03675-5

Cover Photograph courtesy of the U.S. Department of Agriculture, Soil Conservation Service.

Printed in the United States of America

Committee on Conservation Needs and Opportunities

M. GORDON WOLMAN (*Chairman*), The Johns Hopkins University
GEORGE W. BAILEY, U.S. Environmental Protection Agency, Athens, Georgia
SANDRA S. BATIE, Virginia Polytechnic Institute and State University
THOMAS E. FENTON, Iowa State University
WILBUR W. FRYE, University of Kentucky
WILFORD R. GARDNER, University of Arizona
GEORGE W. LANGDALE, U.S. Department of Agriculture, Watkinsville, Georgia
WILLIAM E. LARSON, University of Minnesota
DONALD K. McCOOL, U.S. Department of Agriculture and Washington State University
FRANCIS J. PIERCE, Michigan State University
PAUL E. ROSENBERRY, Iowa State University
LEO M. WALSH, University of Wisconsin

Staff

CHARLES M. BENBROOK, *Project Officer*
CARLA CARLSON, *Editor*

Board on Agriculture

Authors

K. ERIC ANDERSON, U.S. Geological Survey
RICHARD W. ARNOLD, Soil Conservation Service, USDA
B. J. BARFIELD, University of Kentucky
NELSON L. BILLS, Cornell University
LEE A. CHRISTENSEN, University of Georgia
MARION CLAWSON, Resources for the Future
R. H. DOWDY, Agricultural Research Service, USDA, and University of Minnesota
BURTON C. ENGLISH, Iowa State University
KLAUS W. FLACH, Soil Conservation Service, USDA
G. R. FOSTER, Purdue University
DALE A. GILLETTE, National Oceanic and Atmospheric Administration
RALPH E. HEIMLICH, Economic Research Service, USDA
CHRIS J. JOHANNSEN, Purdue University
WILLIAM E. LARSON, University of Minnesota
J. C. McBURNIE, University of Kentucky
ARNOLD R. MILLER, Farm Credit Administration
JOHN A. MIRANOWSKI, Economic Research Service, USDA
WILLIAM C. MOLDENHAUER, Purdue University
CLAYTON W. OGG, Economic Research Service, USDA
RONALD B. OUTEN, U.S. Senate Committee on Environment and Public Works
F. J. PIERCE, Michigan State University
KENNETH G. RENARD, Agricultural Research Service, USDA
GLAUCIO ROLOFF, University of Minnesota
C. FORD RUNGE, University of Minnesota
PAUL E. ROSENBERRY, Iowa State University
R. NEIL SAMPSON, American Forestry Association
MAX SCHNEPF, Soil Conservation Society of America
WESLEY D. SEITZ, University of Illinois
DAVID J. WALKER, University of Idaho
DOUGLAS L. YOUNG, Washington State University

Preface

The papers in this book provide new information on several aspects of soil erosion and applications of the National Resources Inventory (NRI). They were commissioned following a planning workshop in July 1984 and were presented in December 1984 during a national convocation, "Physical Dimensions of the Erosion Problem." The workshop and convocation were held by the National Research Council's Board on Agriculture in response to a request from the Soil Conservation Service of the U.S. Department of Agriculture to facilitate the establishment of discussion between the SCS and natural resource experts. The Board on Agriculture was specifically asked to evaluate the potential applications of the 1982 NRI.

The data provided by these 11 papers and 12 discussions support and expand on the information and conclusions presented in *Soil Conservation: Assessing the National Resources Inventory, Volume 1,* the report of the board's Committee on Conservation Needs and Opportunities.

The papers address three aspects of the NRI: analytical results and methods, specific applications, and resource policy and decision making. New results and methods are described in papers dealing with an improved soil erosion classification scheme, soil erosion productivity damage, and field estimates of C factors. Other papers discuss specific applications of NRI data to ephemeral gully erosion, wind erosion, erosion on rangeland and forestland, erosion control practices, and offsite erosion damage. Resource policy is discussed as it relates specifically to new cropland conversions, targeting soil conservation programs, and use of the NRI in state and local decision making.

The committee is indebted to all who participated in the workshop and convocation, whether as presenters or as additional sources of ideas and information. We particularly appreciate the work of the authors of the technical papers and the discussion papers in writing and revising their manuscripts. These papers make important contributions to the knowledge needed by investigators, analysts, and policymakers to ensure effective use of the NRI and ultimately for control of the process of soil erosion.

M. Gordon Wolman
Chairman

Contents

SOIL CONSERVATION

ASSESSING THE NATIONAL RESOURCES INVENTORY

Volume 2

1

An Improved Soil Erosion Classification: Update, Comparison, and Extension

Ralph E. Heimlich and *Nelson L. Bills*

Problem definition is a critical step in the design and implementation of public policies and programs. Alternate definitions often lead policymakers to a different impression of problem scope or severity and promote conflicting views on viable program options. This type of confusion is clearly evident in the current debate over federal soil conservation policy for U.S. cropland. Policymakers are groping for a precise definition of erodible soil that requires public action. Unless clarified, definitional problems can severely hamper congressional discussions about new soil conservation initiatives.

Studies recently completed by the Economic Research Service (ERS) and the American Farmland Trust (AFT) incorporate new proposals for more systematically classifying land according to its susceptibility to erosion. The proposed classifications help sharpen perceptions of erosion problems. They were developed independently but bear close kinship to one another; each is based on the notion of partitioning the Universal Soil Loss Equation (USLE), which is a predictive model for rainfall erosion, into its physical and management components. This approach offers more precision than the traditional Land Capability Class System (LCCS) or classifications based on annual soil loss rates.

The purpose of this paper is to (1) update the ERS classification system, (2) compare and critique the updated results with those obtained under the AFT proposal, and (3) outline needed extensions and refinements. Recently available sample point information from the 1982 National Resources Inventory (NRI) is used. These data, in conjunction with the 1977 NRI, improve previous efforts to fashion useful descriptions of erosion problems on cropland and to assess trends in the utilization and management of soil resources. The successive inventories allow an evaluation of 1977-1982 changes in land use and land management. Comprehensive 1982 wind erosion estimates may allow wind-induced erosion to be added to these classification systems.

EROSION ASSESSMENT: AN UPDATE TO 1982

Recent research has focused on an erosion classification based on physical erosion potential and the observed range in land management used by farmers in relation to commonly accepted soil loss tolerances (Bills and Heimlich, 1984). Briefly, the product RKLS, a component of the USLE, was used as a measure of physical erosion potential. Cropland was assigned to one of four erosion classes based on RKLS and a 5 tons/acre/year soil loss tolerance, as shown in Table 1. Management options are reflected in the RKLS limits established for each erosion class; limits were derived by dividing the tolerance value by the maximum and minimum CP combination (the management factors) observed in the 1977 NRI. The classification allows identification of soils that are nonerodible because they will not erode above 5 tons/acre/year regardless of management. [The term erodible is now being used instead of the term erosive in line with Sampson's point (1984) that water and wind are erosive but that land can only be erodible.] Highly erodible soils will erode above this limit under even the most stringent conservation management and probably require permanent vegetative cover to control erosion. The remainder, moderately erodible land, is further subdivided according to whether the management actually applied does or does not meet the 5 tons/acre/year goal.

This classification scheme was compared with erosion rate classes and the LCCS using 1977 NRI sample point data (Heimlich and Bills, 1984). The classification was more useful than the alternatives, it was argued, because it separated the contributions that physical and management factors make to erosion. It identifies resources having tractable erosion problems with more precision than the often-used land capability class.

This scheme has now been updated by applying it to sample point data from the 1982 NRI. Results contrasting erosion rates from sheet and rill erosion for 1977 and 1982 are shown in Table 2. For the United States,

TABLE 1 Taxonomy of Cropland Erodibility

Erosion Class	Definition
Nonerodible	RKLS $\leq$ 7
Moderately erodible	
Managed below tolerance	7 < RKLS < 50; USLE $\leq$ 5
Managed above tolerance	7 < RKLS < 50; USLE > 5
Highly erodible	RKLS $\geq$ 50; USLE > 5

SOURCE: Bills and Heimlich (1984).

excluding Alaska, NRI data show that cropland acreage increased about 7 million acres (1.5 percent) between 1977 and 1982.[1] This result is consistent with acreage increases reported by ERS and with data from other sources (Frey and Hexem, 1985; Hexem and Anderson, 1984). ERS cropland estimates, based largely on Census of Agriculture data, show a cropland increase of 12 million acres between 1978 and 1982; harvested cropland increased even more--by 20 million acres.

On balance, changes in land use, conservation management, and data collection procedures between the two inventories have done little to change the erosion status reported earlier. The distributions of cropland across erosion classes as well as erosion rate classes in 1982 are not statistically different from those for 1977.[2] This is not surprising: Five years is too short a period to observe major aggregate adjustments in resource use.

Differences between cropland acreages in the erosion classes range from 1.9 million to 7.8 million acres and may not be large enough to be statistically significant. If the shifts among erosion categories are taken at face value, however, the net increase in cropland over these 5 years was accomplished by substituting nonerodible for highly erodible land. Abandonment or improved management of moderately erodible land losing above 5 tons/acre/year also appears to have occurred. From the standpoint of sheet and rill erosion rates, less cropland eroded at rates between 5 and 13 and over 25 tons/acre/year in 1982 than in 1977. This is consistent with the decrease in the published average annual sheet and rill erosion rates

TABLE 2 U.S. Cropland by Soil Erosion Class and Annual Sheet and Rill Erosion Rate, 1977 and 1982

Annual Erosion Rate[a] (TAY[b])	Non-erodible	Moderately Erodible <5 TAY	Moderately Erodible >5 TAY	Highly Erodible	Total
		(1,000 acres)			
<5					
1977	157,342	161,058	--	--	318,400
1982	165,136	163,626	--	--	328,762
5 to 13					
1977	--	--	56,990	11,150	68,140
1982	--	--	54,988	10,026	65,014
14 to 24					
1977	--	--	5,782	8,672	14,454
1982	--	--	5,872	8,809	14,681
≥25					
1977	--	--	124	11,852	11,976
1982	--	--	85	10,905	10,990
Total[c]					
1977	157,342	161,058	62,896	31,674	412,970
1982	165,136	163,626	60,945	29,740	419,447
		(Percent[e])			
<5					
1977	38.1	39.0	--	--	77.1
1982	39.4	39.0	--	--	78.4
5 to 13					
1977	--	--	13.8	2.7	16.5
1982	--	--	13.1	2.4	15.5
14 to 24					
1977	--	--	1.4	2.1	3.5
1982	--	--	1.4	2.1	3.5
≥25					
1977	--	--	d	2.9	2.9
1982	--	--	d	2.6	2.6
Total					
1977	38.1	39.0	15.2	7.7	100.0
1982	39.4	39.0	14.5	7.1	100.0

[a]Sheet and rill erosion only.
[b]Tons/acre/year.
[c]Detail does not add to published totals due to subsequent coding of some pastureland.
[d]Less than 0.1 percent.
[e]Each entry calculated as percentage of total cropland in 1977 and 1982.

SOURCE: 1977 and 1982 NRIs.

TABLE 3 Percentage Distribution of U.S. Cropland by Soil Erosion Class and Capability Class, Subclass e, 1977 and 1982

Capability Class and Subclass e		Non-erodible	Moderately Erodible		Highly Erodible	Total
			<5 TAY[a]	>5 TAY		
IIe	1977	4.6	10.1	5.5	1.3	21.5
	1982	4.9	10.2	5.6	1.0	21.7
IIIe	1977	5.8	7.1	2.9	3.5	19.3
	1982	5.9	7.1	3.4	3.2	19.6
IVe	1977	2.0	2.6	0.9	1.6	7.1
	1982	2.2	2.6	1.1	1.8	7.7
VIe	1977	0.5	0.8	0.3	0.6	2.2
	1982	0.4	0.8	0.4	0.8	2.4
VIIe	1977	0.1	0.1	0.0	0.2	0.4
	1982	b	0.1	b	0.1	0.3
Subtotal	1977	13.0	20.7	9.6	7.2	50.5
	1982	13.5	20.8	10.5	6.9	51.7
Other sub-classes	1977	25.1	18.3	5.6	0.5	49.5
	1982	25.9	18.2	4.0	0.2	48.3
Total	1977	38.1	39.0	15.2	7.7	100.0
	1982	39.4	39.0	14.5	7.1	100.0

[a]Tons/acre/year.
[b]Less than 0.1 percent.

SOURCE: 1977 and 1982 NRIs.

on cropland from 4.7 tons/acre/year in 1977 to 4.4 in 1982--an expected result, given the increase in non-erodible land and the decrease in highly erodible land in the cropland base.

A different picture emerges when capability classes in the erodible subclass e are compared with the erosion classes developed here (see Table 3). Again, the distributions across subclasses in 1982 are not statistically different from those observed in 1977.[3] Taken literally, cropland in erodible subclasses increased by roughly 8 million acres between 1977 and 1982. The inconsistency between an increase in cropland in the LCCS erodible subclasses and a decrease in cropland in erodible categories of this classification is apparent.

The 1982 NRI data also preserve the inconsistency between the LCCS subclass e and the alternative classification of highly erodible cropland first shown with 1977 NRI data. About 60 percent of highly erodible and moderately erodible cropland managed above tolerance is in low erosion hazard classes IIe and IIIe. Subclasses other than e account for almost a third of moderately erodible cropland managed above tolerance. Conversely, more than one-third of nonerodible cropland is in e subclasses. This is about twice as much land as is in the highly erodible class. Land in LCCS classes with high erosion hazard (IVe through VIIe) appears in all four erosion classes.

In the absence of any systematic error in assigning land capability class ratings or in estimating USLE parameters, two factors could account for the inconsistencies observed between the classification here and the LCCS. First, wind erosion is not included in this classification system but is reflected in the capability class ratings. Thus, some cropland designated nonerodible from a sheet and rill erosion perspective is subject to wind erosion (a point discussed later).

Second, there is no strong conceptual basis for close linkages between the LCCS and soil erosion generated by the interaction of physical and management factors. Subclass e identifies only those soils for which erosion is the dominant limitation. Soils with other limitations can also have substantial erosion, but they are not designated in subclass e. The system is interpretive and applied locally, so it is subject to the judgment of individual technicians and is therefore not consistent.

The LCCS was designed to group "arable soils... according to their potentialities and limitations for sustained production of the common cultivated crops" (Klingebiel and Montgomery, 1961). Its main flaw, according to AFT, was that "its classifications do not reflect modern, scientific estimates of soil erosion rates" (AFT, 1984). It was not intended to distinguish precisely between land resources with differing erosion potential, although it has been used in that fashion in recent years.

In summary, the NRI evidence for 1977 and 1982 shows that sheet and rill erosion decreased from 1.9 billion to 1.8 billion tons/year in the face of an overall increase in acreage cropped. The absolute and relative importance of highly erodible cropland in U.S. agriculture decreased, and conservation management on moderately erodible cropland may have improved slightly. However, more cropland designated as subclass e was in production in 1982 than in 1977, and subclass e was a larger percentage of total cropland. The imprecision with which the subclass e

TABLE 4 American Farmland Trust Taxonomy of Cropland Erodibility

AFT Land Group	Definition
1 Not threatened	RKLS $\leq$ 15
2 Moderately erodible	15 < RKLS < 75
3 Highly erodible	RKLS $\geq$ 75

SOURCE: AFT (1984).

designation identifies land with high erosion potential was as much in evidence in 1982 as it was 5 years earlier.

COMPARISON OF RKLS-BASED CLASSIFICATION SYSTEMS

Although it is clear that a quantitative classification based on RKLS is desirable, the exact classifying criteria are not obvious. Different RKLS-based classifications serve different purposes and their advantages vary with the purpose they are addressing. In this section, two RKLS-based classifications are compared using 1982 NRI cropland data for the United States.

The AFT, in a comprehensive review of soil conservation policy, recommended that "cropland in the U.S....be designated into one of three groups by local conservation districts on the basis of practical, consistent, and scientifically sound criteria reflecting the land's vulnerability to erosion" (AFT, 1984). The Trust proposed a classification based on the land's inherent erosion potential, as measured by RKLS. The definition of each group is shown in Table 4. Group 1 land, under AFT's system, is not threatened by erosion and is capable of sustaining continuous, intensive agricultural use. It may have other conservation problems, however, such as drainage or salinity, and may not be the best land in terms of current yield potential. Group 2 land is moderately erodible and is envisioned as the focus of USDA's (U.S. Department of Agriculture) traditional conservation programs and practices. Group 3 consists of highly erodible land for which conversion to permanent cover is probably the most cost-effective means of

TABLE 5 U.S. Cropland by Soil Erosion Class and AFT RKLS Groups, 1982

AFT RKLS Groups[a]	Non-erodible	Moderately Erodible	Highly Erodible	Total
		(1,000 acres)		
Group 1	165,136	114,756	--	279,892
Group 2	--	103,303	12,253	115,556
Group 3	--	6,512	17,487	23,999
Total[b]	165,136	224,571	29,740	419,447
		(Percent[c])		
Group 1	59.0	41.0	--	100.0
Group 2	--	89.4	10.6	100.0
Group 3	--	27.1	72.9	100.0
Total	39.4	53.5	7.1	100.0

[a]Groups 1, 2, and 3 are RKLS $\leq$ 15, 15 < RKLS < 75, and RKLS $\geq$ 75, respectively (see Table 4).
[b]Detail does not add to published totals due to subsequent coding of pastureland.
[c]Calculated as percentage of each erosion class.

SOURCE: AFT (1984) and 1982 NRI.

reducing erosion. A long-term conservation reserve is proposed by AFT as the primary means for encouraging such conversion on group 3 land.

The ranges of RKLS in AFT's system result from applying "normal farming conditions" to achieve specified ranges of erosion rates without traditional conservation practices. Thus, under average management (C factor = 0.30), an RKLS of up to 15 yields sheet and rill erosion of less than 5 tons/acre/year. The second group would have erosion rates of less than 15 tons/acre/year under normal farming conditions, which could be corrected using traditional conservation practices. The third group, barring extraordinary and very costly conservation systems, could not produce cultivated crops without eroding above 15 tons/acre/year.

Although the AFT classification is similar to the RKLS-based system described in this paper, a different picture of erosion problems emerges (see Table 5). Under AFT, two-thirds of U.S. cropland is not considered threatened by erosion. Under the classification detailed earlier, more than 41.0 percent of this land is moderately erodible because soil loss above 5 tons/acre/year is expected if the level of conservation management applied to it is below average. Conversely, 10.6 percent of the cropland in AFT's group 2 would erode above a tolerable level except under the most restrictive combinations of crop rotation, tillage, and conservation practices.

The AFT report also discussed tactics that might be used by USDA to implement a new system. The authors suggest that primary technical responsibility for designating cropland into the three groups should rest with local conservation districts. Guidelines for such a grouping would be developed at the national or state level. AFT recommends grouping land by capability class and subclass as an interim measure "until a superior system can be developed or major flaws in the existing capability classifications can be corrected" (AFT, 1984).

Unfortunately, AFT's interim groups do little to overcome the difficulties seen above. Comparing AFT's interim groups to the classification here reveals that group 1, which is not supposed to be threatened by erosion, contains almost as much moderately erodible land as nonerodible land. The "moderately erodible" group 2 contains more of what is classified here as highly erodible land than does the "highly erodible" group 3. There is more nonerodible land in group 3 than there is highly erodible land. Thus, the proposed interim grouping has little to recommend it as a way to distinguish cropland resources requiring different kinds of conservation management because lands of all kinds are present in each group.

Differences between the two systems can be traced to the way management is represented in the calculations. This paper considers the combined effects of cropping system (including rotations, residue management, and tillage) and conservation management (including traditional conservation practices such as contour plowing, stripcropping, and terraces). The AFT system considers only cropping system and assumes that conservation practices can be applied in the future, even if they are not now present. The RKLS limits specified in this paper are based on the best and worst combinations of management factors observed, while AFT's are based on the "normal farming conditions" alone.

The purpose, in both systems, is to measure the physical potential for erosion, abstracting from the management currently applied. The object is to determine if the resource can or cannot meet a soil loss goal within the relatively fixed physical constraints imposed by climate, topography, and soil type. However, the system proposed here is based on the premise that both cropping system and conservation management are free to range over the entire spectrum of technology currently available. Development of conservation tillage systems has tended to further blur the distinctions between practices undertaken to produce a crop and practices used to retard erosion. Only the product CP reflects all the short-term management practices to control erosion over which operators have some discretion. Considering both cropping and conservation management at the extremes of their practical range, which this paper does, provides a more accurate picture of resource capability with respect to erosion.

Despite differences in specification, the similarities between these independently derived classifications are important. Two basic themes are present in both. Each incorporates an objective, scientific, quantitative measure of physical erosion potential, separate from the management currently applied to the land. Both utilize the concept of triage, borrowed from medical practice, in which three groups are defined: land that needs no erosion treatment because it has no erosion potential; land with so much erosion potential that no treatment will reduce erosion to acceptable levels; and the remaining land for which treatment is needed and will reduce erosion to acceptable levels. These basic themes set the RKLS-based classifications reviewed here apart from the LCCS.

NEEDED EXTENSIONS

The RKLS-based classification systems discussed above have two shortcomings that need to be addressed. First, estimates for the entire United States in the 1982 NRI showed substantial wind erosion, which makes a classification based solely on sheet and rill erosion questionable. Wind erosion is estimated using an empirical equation (analogous to the USLE) developed in the mid-1960s by Woodruff and Siddoway and made operational by Skidmore and Woodruff (1968). In the 1977 NRI, these estimates were

TABLE 6 U.S. Cropland by Soil Erosion Class and Annual Wind Erosion Rate, 1982

Annual Wind Erosion (TAY)[a]	Non-erodible	Moderately Erodible		Highly Erodible	Total
		<5 TAY[a]	>5 TAY		
		(1,000 acres)			
<5	123,979	148,164	54,288	28,736	355,167
5 to 13	28,117	10,889	4,306	811	44,123
14 to 24 7,010	2,888	1,294	109	11,301	
≥25	6,030	1,057	2,143	84	8,856
Total[b]	165,136	163,626	60,945	29,740	419,447
		(Percent[d])			
<5	29.6	35.3	12.9	6.9	84.7
5 to 13	6.7	2.6	1.0	0.2	10.5
14 to 24	1.7	0.7	0.3	d	2.7
≥25	1.4	0.4	0.3	d	2.1
Total	39.4	39.0	14.5	7.1	100.0

[a]Tons/acre/year.
[b]Detail does not add to published total due to subsequent coding of pastureland.
[c]Each entry calculated as percentage of total cropland.
[d]Less than 0.1 percent.

SOURCE: 1982 NRI.

confined to the Great Plains states. Wind erosion estimates now available for the United States from the 1982 NRI show that 40 percent of all erosion on cropland is from wind; in 12 states wind causes more than half the cropland erosion. The magnitude of wind erosion has serious implications for RKLS-based erosion classifications. Some cropland in arid regions may be prone to high wind erosion but appears in the nonerodible category of the classification used here. In areas where both wind and water erosion occur, cropland that is moderately erodible from a sheet and rill perspective may erode above tolerance level when the effects of wind and water are combined.

The impact of wind erosion on the classification developed earlier can be seen by arraying wind erosion rates against the RKLS-based erosion classes (*see* Tables 5 and 6). Almost 85 percent of all cropland has wind erosion rates below 5 tons/acre/year. Only a quarter of the cropland rated nonerodible from a rainfall standpoint has wind erosion rates above 5 tons/acre/year while less than a tenth of moderately erodible cropland managed below 5 tons/acre/year has wind erosion in excess of that level. Thus, consideration of wind erosion shifts 41.1 million acres of cropland out of the nonerodible class and 14.8 million acres of well-managed moderately erodible cropland to the erosively managed category. Small acreages might also be shifted to the highly erodible category based on wind erosion.

To overcome this difficulty, a wind erosion classification should be developed analogous to this RKLS classification. Some problems are introduced because the Wind Erosion Equation, unlike the USLE, is not simple multiplicative relationship. Skidmore and Woodruff's (1968) equation is given by:

$$E = IKCf(L)\ f(V)$$

where E = potential average annual soil loss in tons/acre/year; I = soil erodibility, based on percentage of soil particles less than 0.84 mm in diameter; K = soil ridge roughness in relation to a 1:4 ridge height to spacing ratio; C = climatic factor, a function of average annual wind speed and the Thornthwaite precipitation-evaporation index; f(L) = a function of field length along the direction of prevailing wind; and f(V) = equivalent vegetative cover, a function of flat small grain residue equivalents.

As with the USLE, it is important to distinguish factors that reflect relatively unchanging physical constraints from those that are subject to annual change by the farmer. Only the climate (C) and soil erodibility (I) factors reflect such physical constraints. Field length [f(L)] is altered by wind breaks and stripcropping. Similarly, soil ridge roughness (K) is affected by the depth and spacing of tillage, and vegetative cover [f(V)] depends on crop rotation and residue management.

Unfortunately, the classification cannot be extended yet because the necessary wind erosion equation parameters are not currently listed on the NRI computer tape. Inclusion of fields with records of soil erodibility (I) and climatic (C) factors used to calculate the wind erosion estimate would allow this wind erosion classification to be made. The wind erosion classification could be displayed separately and also combined with results from the USLE classification to show a more complete picture of erodible cropland resources.

A second shortcoming of RKLS-based erosion classifications is the reliance on soil loss tolerances. The system used here was defined in relation to a single 5 tons/acre/year soil loss tolerance goal, while the AFT groups were referenced generally to existing tolerance values ranging from 2 to 5 tons/acre/year (AFT, 1984; Bills and Heimlich, 1984). The choice of a tolerable soil loss goal is of more than academic interest because of the policy implications of the erosion classification schemes proposed. Setting goals too low forces large acreages into the highly erodible category, in which erosion control is largely synonymous with conversion to permanent cover.

For example, under a 5 tons/acre/year goal, about a third of the cropland in such productive regions as the Iowa and Missouri deep loess hills [Major Land Resource Area (MLRA) 107] and the Palouse and Nez Perce plains (MLRA 9) falls into the highly erodible category. It is not clear that erosion rates above 5 tons/acre/year will reduce the long-term productivity of the deep soils in such areas. Thus, it is equally unclear that cropland in such areas should be classed "highly erodible."

Conversely, setting soil loss goals too high forces acreage that might suffer erosion damage into the nonerodible category. Under a 5 tons/acre/year goal, almost 20 percent of cropland in the Northeast is considered nonerodible, while under actual T values (soil loss tolerance limit) assigned to each soil, only 10 percent of the cropland in this region of shallow soils falls in the nonerodible category.

It is important to note that the problem lies not in the classification scheme, but in the tolerable soil loss goals adopted. Recently, existing assigned T values have been the subject of a great deal of criticism and scrutiny. Briefly, critics of existing T values claim that they are either too low, overprotecting deep soils that would suffer no loss of productivity at higher erosion rates (Cook, 1982), or that they are too high to reflect actual soil formation rates from parent material (OTA, 1982). At the center of the controversy are new models of the soil erosion/soil productivity relationship, such as EPIC (Erosion Productivity Impact Calculator) and the Minnesota model (Pierce et al., 1983; Williams et al., 1983). Results from these new models indicate that maintenance of long-term productivity is possible with a wider range of soil loss rates than the existing T values of 2 to 5 tons/acre/year.

The results from objective erosion/productivity models can eventually be substituted for the existing, subjective T factors. For example, soil loss tolerances could be based on explicit losses in productivity judged acceptable over a specific planning horizon. This has already been done by Pierce et al. (1984) for soils in Dakota County, Minnesota, using a 5 percent allowable decline in yield over 100 years.[4] Their Tl, based on inherent soil productivity, ranged from 1.3 to 40 tons/acre/year, a much broader range than conventional T factors. If such values could be computed for all soils in agricultural use and associated with the NRI sample records through the Soils-5 identification field on the record, a classification of cropland according to inherent productivity could be produced. If many of the values nationwide are higher than the existing 5 tons/acre/year maximum T factor, as they are in Dakota County, the proportion of cropland in the highly erodible and erosively managed categories will be even smaller than it is now.

SUMMARY AND CONCLUSIONS

This paper addresses the role that problem definition plays in the design and implementation of public soil conservation policy. Point sample data from the 1982 NRI were used to update an RKLS-based soil erodibility classification recently proposed by the ERS. The system was compared with a similar proposal by the AFT and contrasted with the traditional LCCS.

Four principal conclusions can be drawn from the discussion. First, the LCCS is flawed when used in quantitative assessments of erosion potential on cropland because it fails to link land capability class-subclass designations with soil loss outcomes produced by the interaction of physical and management factors. This is not an indictment of the LCCS itself, but a reflection of the tendency to use it for tasks it was not designed to accomplish. LCCS was devised long ago to classify soils according to the type and severity of hazards encountered when land is used to grow commonly cultivated crops. However, the evidence presented here clearly demonstrates the system's inadequacy for assessing soil erodibility of cropland. Such assessments are critical because the prospect for more incisive public policy is tied to improved incentives for changes in conservation management on cropland with the physical characteristics to benefit from it.

Second, efforts to devise alternative RKLS-based classifications seem promising. They are already used for planning purposes at the farm level, are tractable, incorporate scientific techniques for estimating soil erodibility, and greatly sharpen the focus of public policy options for mitigating soil erosion. However, the results obtained with RKLS-based classifications are sensitive to the conventions used to deal with land management. The specification of highly erodible land depends on the level of conservation management it is reasonable to expect farmers to use. If soil loss tolerances cannot be achieved with feasible cropping systems, land should be taken out of crop production.

Third, wind erosion--now recognized as an important cause of soil loss--is not encompassed in existing systems but clearly needs to be. The same principles of separating physical and management alternatives should be appropriate for wind erosion. An analogous classification based on the Wind Erosion Equation seems feasible but cannot be empirically tested until the equation parameters collected in the 1982 NRI are available.

Finally, this experimentation with new classification systems underscores the importance of soil loss tolerances in the continuing debate over soil conservation policy. The proposed system is flexible insofar as any set of soil-specific tolerances are readily accommodated. Unless the soil loss goals specified are scientifically based and accurately express the social significance of continued soil loss on cropland, however, large acreages can be misclassified. The nation can afford neither the immediate loss in production caused by mistakenly withdrawing land from cultivation nor the future loss in productivity from continuing to grow crops on excessively erodible land.

NOTES

1. The 1982 cropland estimate of 419 million acres is about 2 million acres less than the amount reported in previous NRI-published summaries. The small discrepancy is due to a reclassification of some sample points from pastureland to cropland.
2. The hypothesis that distributions of cropland were identical in 1977 and 1982 was tested using a chi-square

statistic calculated as the sum of squared differences between the 1977 and 1982 proportions, divided by the 1977 proportion. Chi-square statistics for distributions of erosion rates and erosion classes, respectively, were 0.1136 and 0.1233, not significantly different at a 95 percent confidence level.

3. The chi-square statistic for the distribution of subclass e cropland across erosion classes in 1977 and 1982 was 0.1586, not significantly different at a 95 percent confidence level.

4. Although the necessary exercise of judgment in setting allowable yield declines for T1 values can be criticized on the same grounds as existing T values, it is more direct and scientific. Judgments on acceptable yield decreases get directly to the matter of long-term productivity loss and are used in conjunction with more scientific estimates of the effect of continued erosion on crop yields.

REFERENCES

AFT (American Farmland Trust). 1984. Soil Conservation in America: What Do We Have to Lose? Washington, D.C.: American Farmland Trust.

Bills, N. L., and R. E. Heimlich. 1984. Assessing Erosion on U.S. Cropland: Land Management & Physical Features. AER-513. Washington, D.C.: Economic Research Service, U.S. Department of Agriculture

Cook, K. 1982. Soil loss: A question of values. J. Soil Water Conserv. 37:89-92.

Frey, H. T., and R. W. Hexem. 1985. Major Uses of Land in the United States: 1982. AER-535. Washington, D.C.: Economic Research Service, U.S. Department of Agriculture.

Heimlich, R. E., and N. L. Bills. 1984. An improved soil erosion classification for conservation policy. J. Soil Water Conserv. 39:261-266.

Hexem, R. W., and W. D. Anderson. 1984. Cropland Use and Supply: Outlook and Situation Report. CUS-1. Washington, D.C.: Economic Research Service, U.S Department of Agriculture.

Klingebiel, A. A., and P. H. Montgomery. 1961. Land Capability Classification. Agriculture Handbook No. 210, USDA Soil Conservation Service. Washington, D.C.: U.S. Government Printing Office.

OTA (Office of Technology Assessment). 1982. Impacts of Technology on U.S. Cropland and Rangeland Productivity. Washington, D.C.: U.S. Government Printing Office.
Pierce, F. J., W. E. Larson, R. H. Dowdy, and W. A. P. Graham. 1983. Productivity of soils: Assessing long-term changes due to erosion. J. Soil and Water Conserv. 38:39-44.
Pierce, F. J., W. E. Larson, and R. H. Dowdy. 1984. Soil loss tolerance: Maintenance of long-term soil productivity. J. Soil Water Conserv. 39:136-138.
Sampson, R. Neil. 1984. Sloppy Semantics (letter). J. Soil Water Conserv. 39:282-283.
Skidmore, E. L., and N. P. Woodruff. 1968. Wind Erosion Forces in the United States and Their Use in Predicting Soil Loss. Agriculture Handbook No. 346, USDA Science and Education Administration. Washington, D.C.: U.S. Government Printing Office.
Williams, J. R., K. G. Renard, and P. T. Dyke. 1983. EPIC: A new method for assessing erosion's effect on soil productivity. J. Soil Water Conserv. 38:381-384.

Discussion

Richard W. Arnold

The paper by Heimlich and Bills illustrates well that each land classification system more or less does its own thing, and that no one system serves as an adequate surrogate for another. It is possible to compare land capability classes with RKLS (physical erosion potential) classes, but not to substitute one for the other. Each has its own purpose, and each its own criteria. The search must continue for ways to identify potentially erodible soils and to establish rational limits of productivity loss in order to distinguish those soils that suffer more productivity damage than is acceptable.

A great deal has been learned about the relationships between soil loss and crop productivity, and refinements in yield models are improving yield predictions. But there are still not enough data supporting these relationships to make national applications of such modeling results. In the meantime, the alternative RKLS-based classifications proposed by Heimlich and Bills seem to be promising. One obvious modification is to normalize or standardize the values, either by dividing them <u>into</u> T values or by dividing them <u>by</u> T values.

When a T value is divided by an RKLS value for a given soil, the quotient (CP value) represents the combination of crop cover and conservation practices (CP in the Universal Soil Loss Equation) that would achieve the assigned T value. The quotient generated decreases as RKLS increases, a relationship between potential erodibility and CP numbers that is often forgotten. In the second approach, where RKLS values are divided by T for given soils, the numbers represent the reciprocal of the CP combinations. But more important, the numbers increase as potential soil losses increase, making the relationship easier to comprehend and use in planning and making decisions on land use.

Not only does the use of T values help to normalize the information, it also carries with it a subjective notion of the importance of soil loss in changing the soil environment and reducing long-term crop production. Where sustainable production is influenced mainly by sheet and rill erosion rather than by wind erosion or by a combination of wind and water erosion, a classification based on RKLS/T appears to be a reasonable compromise and one that is feasible to implement at the field level. Potential erodibility classes that parallel the triage proposed by Heimlich and Bills would be as follows:

Erodibility Class	RKLS/T	USLE Estimated Soil Loss
Nonerodible soils	<1.4	<T
Moderately erodible soils (capable of achieving T with current practices)	1.4-10	a) <T (protected) b) >T (needing protection)
Highly erodible soils (not capable of achieving T with current practices)	>10	>T

The RKLS/T values can also be thought of as representing the potential annual rate of sheet and rill erosion per unit of tolerable limits, that is, the tons of soil loss per unit of T without crop cover or conservation practice. For the classes that parallel those of Heimlich and Bills, those limits again would be 1.4 and 10.

A review of such an RKLS-based system is being done by the RCA Fragile Soils Work Group, a USDA interagency committee, using the 1982 NRI data. Preliminary estimates indicate that in 1982, about 37 percent of the cropland was nonerodible. About 51 percent was moderately erodible, and the remaining 12 percent--some 53 million acres of cropland--was highly erodible. Perhaps up to one-third of the highly erodible land may someday be considered moderately erodible or at least controllable if improved crop cover and conservation practices are developed and adopted by farmers. This shift could occur if the best CP combination values were lower than at present.

Discussion

K. Eric Anderson

The paper by Heimlich and Bills should be welcomed for reopening the whole question of properly defining the problem of soil erosion and asking which classification of data is appropriate for that problem. That is always an important issue in research activities. These remarks focus on the types of technology that are beginning to emerge that will be relevant to the conduct of the next National Resources Inventory (NRI), whether the inventory is focused solely on questions of erosion or on other issues of land management.

The Geological Survey is moving very rapidly in the direction of computer data bases from map information. Data bases are beginning to emerge that include large quantities of information from topographic maps, such as elevation, transportation, and hydrography. These are going to offer some very new opportunities in the measurement of slope, aspect, and drainage, which will contribute substantially to studies of erosion and to allowing users to begin to address--both in a fairly local way and with broad regional perspective--detailed, comprehensive studies of the physiographic character of the land that will help advance many of these studies.

There are a variety of applications of these data bases. For example, the Geological Survey is deeply involved in conducting the 1990 census of population whereby a data base covering the entire United States will be completed to support the conduct of that census. There are some parallel potential applications within the field of agriculture. The agency is working closely with parts of the Soil Conservation Service to develop technology for the construction of data bases on soils information.

A whole new technology is emerging in terms of geographic information systems that will allow users to combine this information, analyze it, and display it rapidly in many different forms. These analytical capabilities will have considerable impact on studies of erosion and land management. In fact, a number of the land management agencies in the United States are actively developing and beginning to apply this geographic information system technology.

The other arena in which technology will soon begin to have substantial impact is remote sensing. The Geological Survey has been involved over the last 3 years in a study of irrigated cropland in the Midwest, the principal concern being the depletion of aquifers because of the withdrawal of water for irrigation. Satellite remote sensing has come up with far better estimates of the actual water consumption than were ever available before.

All these technologies--in terms of information systems and new ways of collecting data from satellites--will have a substantial impact on the whole NRI effort in the future.

2

Assessing Soil Erosion Productivity Damage

David J. Walker and *Douglas L. Young*

Scientists have long recognized in a general sense that erosion reduces crop yields; however, economic assessment of erosion damage has always been elusive. The erosion process is gradual, and annual yield variability from weather, disease, and pests obscures the inexorable reduction of crop yields from soil loss. In many regions, technical progress has boosted crop yields faster than erosion has reduced them, perhaps persuading some that erosion damage is of no consequence. Yet, technology may only be masking erosion damage in many instances. In fact, some types of technical progress can actually increase erosion damage. A correct assessment requires that the effects of erosion and technology on yields be disaggregated and that separate projections be made of their impacts.

Accurate measures of erosion damage are important. Many farmers, while admitting that erosion may reduce crop yields and farm income in the future, may be unwilling to adopt conservation with only a vague notion of the long-term cost of erosion damage. The Soil Conservation Service's (SCS) recent targeting of conservation efforts stresses the need for information on erosion damage. Rather than identifying areas for action on the basis of high erosion rates, target areas could be selected based on net economic benefit of conservation, including the erosion damage avoided. Simple physical criteria, such as soil loss, may be misleading because some areas with high erosion rates may not suffer reduced yields if subsoils are deep and suitable for cultivation. Even when yield loss is significant, if the crops have a low value, the economic loss may not be as great as in an area with moderate erosion but high-value crops.

OBJECTIVES AND SCOPE OF ANALYSIS

Damage from soil erosion is generally divided into onsite productivity impacts and offsite environmental effects. Although both components are necessary to compute aggregate economic damage of soil erosion, this paper focuses exclusively on the measurement of onsite productivity effects, in light of the authors' conceptual and empirical research specialization. This focus does not reflect a belief that economic measurements of offsite environmental impacts are not equally important.

The objectives of this paper are to: (1) develop and explain fundamental concepts for the correct assessment of productivity damage from soil erosion, (2) present empirical evidence from the Pacific Northwest Palouse region on the historical nature of technical progress and its interaction with erosion as both processes have influenced winter wheat yields through time, (3) discuss an explicit wheat yield projection model for the Palouse based on historical patterns and rates of technical progress and erosion interactions, (4) present an operational computerized erosion damage assessment model that incorporates the Palouse wheat yield projection model, (5) discuss possible uses of the National Resources Inventory (NRI) data for regional and national erosion damage assessments, and (6) summarize the research and policy implications of the analysis.

CONCEPTS FOR MEASURING EROSION DAMAGE

There are four concepts involved in the assessment[1] of erosion damage: (1) a basic comparison of yields with and without conservation (or with and without erosion), (2) an awareness that the yield penalty from using conservation tillage should not confound the assessment of erosion damage, (3) the identification of residual and reparable yield damage, and (4) the need to separate the effects of technological change from those of erosion.

The first three concepts apply to erosion damage measurement either with or without technical progress, whereas the fourth concept explicitly considers the influence of concurrent technical progress on erosion damage measurement. A physical measure of erosion damage--yield damage--is used in this section to illustrate the concepts for measuring erosion damage, and then an economic measure of erosion damage is presented based on those concepts.

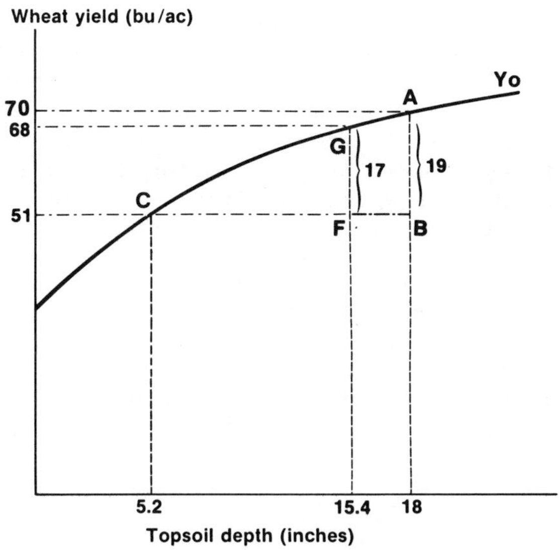

FIGURE 1 Yield damage with zero-erosion basis and conservation basis.

Compare With Versus Without

The basic idea underlying the measurement of erosion damage is the "with versus without" comparison so common in economic analysis. Two possible comparisons are relevant. In a comparison of yield with erosion versus yield without, the basis for comparison is yield after zero erosion, i.e., with unchanged topsoil depth. Erosion damage is the lost yield from gross erosion with conventional tillage. Alternatively, the erosion damage comparison could be yield with conservation versus yield without. Here, the basis for comparison is dynamic or changing over time, yield with topsoil depth conserved using the most cost-effective conservation tillage system available. Erosion damage is the lost yield from the additional erosion under the conventional (erosive) tillage system compared to conservation tillage erosion.

These two bases for measuring erosion damage can be illustrated graphically (see Figure 1). Consider, first, erosion damage with a zero erosion basis for comparison. The yield with a current topsoil depth of 18 inches is 70 bushels/acre. After 64 years of erosion with conventional erosive tillage, topsoil declines to 5.2 inches and yield declines to 51 bushels/acre. Erosion damage with this measure is 19 bushels/acre, the difference between yield with eroded soil and yield without erosion. With the best conservation alternative as the basis for comparison, topsoil depth and yield after 64 years is 15.4 inches and 68 bushels/acre, respectively. Erosion damage with this measure is 17 bushels/acre, the difference between yield with eroded soil and yield with conserved soil.

There are instances when either measure may be more appropriate. The measure derived from the zero erosion basis may be more useful for evaluating alternative conservation practices in a region, while that derived from the dynamic basis, or comparison with the best conservation alternative, might be more useful for selecting target areas for conservation emphasis.

Avoid Confounding Tillage Yield Penalty and Damage

Even with a dynamic basis for comparison, where two different relevant tillage systems generate the two topsoil depths used for the yield damage assessment, one yield-topsoil depth response function must be used to measure yield at both the conserved and eroded topsoil depths. The conservation tillage response function should be used on the premise that ultimately conservation will be required to protect the soil.

It would be a mistake to use separate yield-topsoil depth response functions for conservation and conventional tillage to estimate yields at the conserved and eroded topsoil depths, respectively. Often the conservation tillage system will yield less at the same topsoil depth than the conventional system, causing its response function to lie below that for the conventional system (<u>see</u> Figure 2). If the yield at the eroded soil depth (5.2 inches) is measured with the conventional yield function (Ye) but the yield at the conserved topsoil depth (15.4 inches) is measured with the conservation yield function (Yc), erosion damage would be confounded with the tillage yield penalty. In cases where yields are lower with conservation tillage, using both yield functions would underestimate the damage attributable to erosion (10 bushels/acre versus 17).

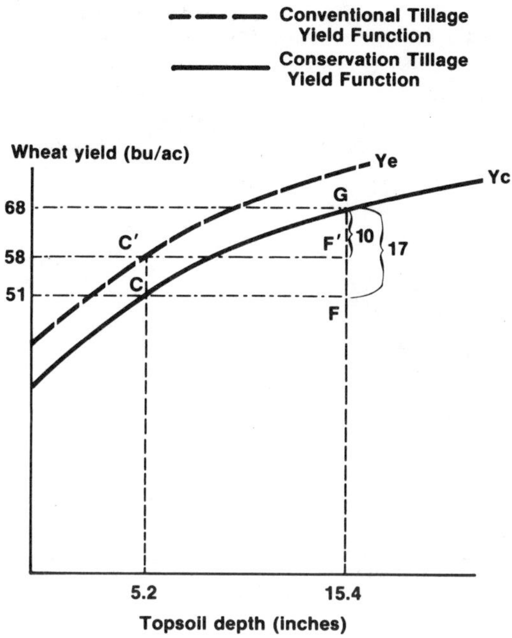

FIGURE 2 Avoid confounding erosion damage with tillage penalty.

Distinguish Between Reparable and Residual Yield Damage

It is useful to partition yield decline from soil erosion into two components, reparable damage and residual damage. Reparable damage is usually associated with loss of soil fertility from erosion and is that portion of the yield decline from erosion that can be restored by increasing organic matter, fertilizer, or other inputs. After economically optimal input adjustments, there will usually be residual yield damage due to deterioration in the soil environment. Reduced moisture infiltration and retention capacity, diminished rooting zone, and impaired soil structure cause residual damage to yields that cannot be remedied economically.

To distinguish between reparable and residual yield damage, consider Figure 3. The restored-yield curve incorporates the yield impact of making economic adjustments in inputs as erosion proceeds, but the constant-input curve reflects yields if no such adjustments are

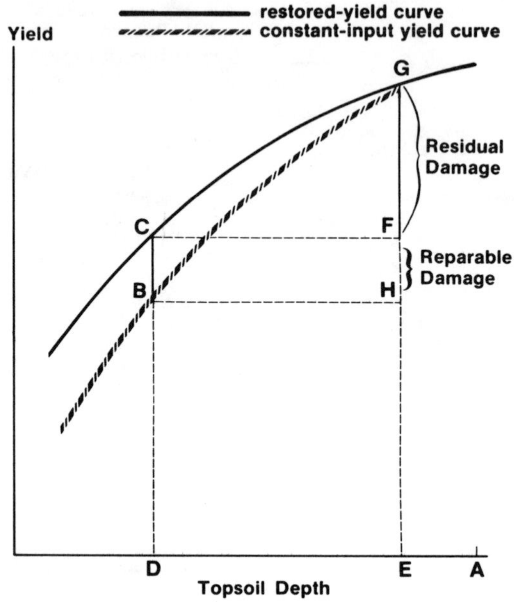

FIGURE 3 Residual and reparable erosion damage.

made. From initial topsoil depth A, using the conservation practice for a number of years would reduce topsoil depth to E and yield to G--providing the basis for comparison with the erosion alternative. If an erosive practice were used for the same number of years, topsoil depth would be reduced further, to D. Yield would decline with erosion, whether economic input adjustments are made (GC) or not (GB). Total yield damage is given by GH.

Some of this yield damage might be restored, depending on subsoil and climatic factors. By increasing fertilizer or other soil-substituting inputs, yield could be restored to point C. The yield decline that can be restored is the reparable component of erosion damage, BC in Figure 3. The remedy must cost less than the value of the yield damage restored. This cost would be included as part of the economic cost of erosion damage.

The restored-yield curve reflects the relationship between yield and topsoil depth after profit-maximizing input adjustments to erosion have been made. Although topsoil depth is the only explanatory variable illustrated in Figure 3, other inputs also vary in the restored-yield function. The extent of the input adjustment process is limited by yield response to increased inputs, by the cost of those inputs, and by output value. The restored-yield curve indicates residual damage to yields. GF in Figure 3 is the difference between yields with conserved soil and yields with eroded soil after profit-maximizing input adjustments have been made. Throughout this paper, reference to yield damage always means this residual yield damage, and yield damage is always measured along the restored-yield curve.

Separate the Effects of Erosion and Technology

Erosion damage assessment is considerably complicated by the impact of technical progress on crop yields. This section first considers technological change that is exogenous with respect to erosion--that is, the rate of technical progress is independent of the rate of erosion. Later, technology that is induced by erosion is considered.

Yield observations over time are confounded by the joint influence of erosion that reduces yield and technology that increases it. In regions where the latter effect predominates, failure to disaggregate this joint influence could lead to the erroneous conclusion that erosion damage does not exist. With exogenous technical progress, erosion damage should not be measured as an absolute decline in historical yield but as the decrease in _potential_ yield with technology and conservation. This requires establishing how much higher yields would be with new technology if soil is conserved. It is necessary to separate the projected effects of erosion and technology. Simply ignoring technology could result in overestimates _or_ underestimates of erosion damage, depending on the interaction between technical progress and topsoil depth.

The effect of exogenous technical progress on erosion damage can be illustrated beginning with Figure 4. Curve Yo illustrates the yield damage from the additional erosion with conventional tillage compared to conservation tillage over a 64-year period with static technology; that is, no technical progress in yields. In this example (identical to Figure 1), erosion with conventional tillage over 64 years would reduce topsoil depth to 5.2 inches and yield to 51 bushels/acre. Using conservation tillage over the same period would reduce topsoil and yield by

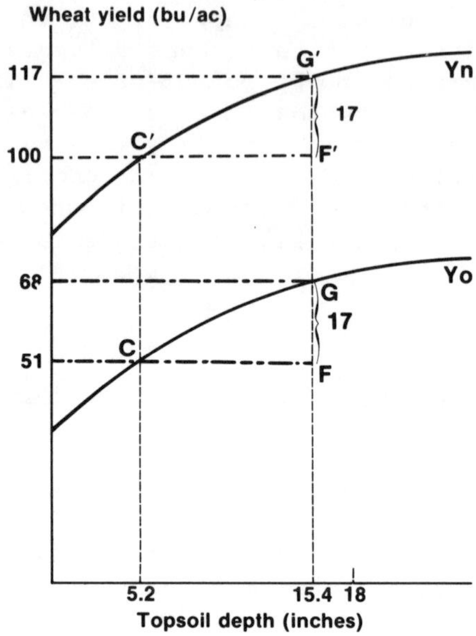

FIGURE 4 Residual yield damage with land-neutral technical change.

less, to 15.4 inches and 68 bushels/acre. The difference, yield with conservation versus yield without conservation, is 17 bushels/acre and measures yield damage in the absence of technical progress.

Land-Neutral Technology

Land-neutral technical progress is illustrated in Figure 4. Technology shifts the yield function upward from Yo to Yn, increasing yield by an equal absolute amount at each topsoil depth. Land-neutral technical progress is most likely on cropland with deep, friable subsoils. In the absence of technology, yield would have declined from G to C. Because technology boosts yield from G to C' in spite of erosion, one might conclude that technology had eliminated erosion damage. That faulty conclusion, however, is based on a "before versus after" erosion comparison which confounds exogenous technology and erosion damage.

A correct measure of erosion damage is based on a with conservation versus without conservation comparison of yield along the technology-augmented yield function, Yn. This measure of erosion damage is the difference between potential yield with conservation and exogenous technology versus realized yield with erosion and the same technology. Potential yield declines from G' at the conserved soil depth to C' at the eroded soil depth giving a yield damage measure of 17 bushels/acre after 64 years of erosion.

Ignoring technology by measuring yield damage along Yo, which assumes static technology, produces an identical measure of yield damage, 17 bushels/acre. Even though there has been an upward yield trend over time in this example, exogenous technology has not reduced erosion damage.

Land-Complementary Technology

Land-complementary technical progress boosts yields more at deeper topsoil depths as illustrated by the shift from Yo to Yn in Figure 5. Improved crop cultivars might be an example of land-complementary technical change. Improved crop varieties usually realize their greatest genetic yield potential in a soil environment with nonrestrictive moisture and nutrient supplies. These conditions are more often found on less eroded sites.

Because land complementary technical change increases the slope of Yn relative to Yo in Figure 5, the appropriate measure of erosion damage, lost potential yield of G'F', is greater than the erosion damage, GF, which would be measured if technology were ignored. Land-complementary technology actually increases erosion damage (32 bushels/acre versus 17 bushels/acre) when damage is measured appropriately.[2]

Land-Substituting Technology

Land-substituting technology boosts the yield function more at shallower topsoil depths as shown in Figure 6. An example might be tillage improvements that conserve soil moisture. Because topsoil serves as a moisture reservoir, moisture deficiency is more likely at eroded sites with shallow topsoil. Thus, technical advances that conserve soil moisture are likely to boost yields more for shallower topsoils. Compared with no technology, yield damage decreases with land-substituting

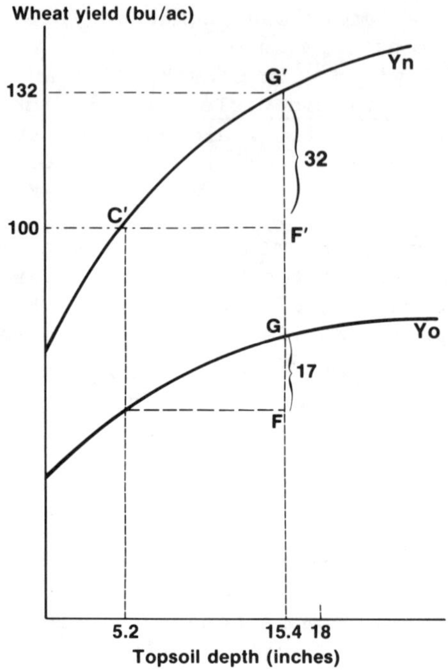

FIGURE 5 Residual yield damage with land-complementary technical change.

technology (11 bushels/acre versus 17 bushels/acre) because the technology-shifted yield function (Yn in Figure 6) becomes less steep.

Because erosion damage with exogenous technology is measured along a single technology-augmented yield function, only the case of land-substituting exogenous technology mitigates erosion damage. This reduction in yield damage is due solely to reduced slope of the yield function and occurs regardless of whether or not there is an upward yield trend over time.

It is important to incorporate technology projections in erosion damage assessment. Ignoring technical progress will result in unbiased damage estimates only with land-neutral technical progress. Ignoring land-complementary technology will underestimate erosion damage. Ignoring land-substituting technical progress will overestimate erosion damage.

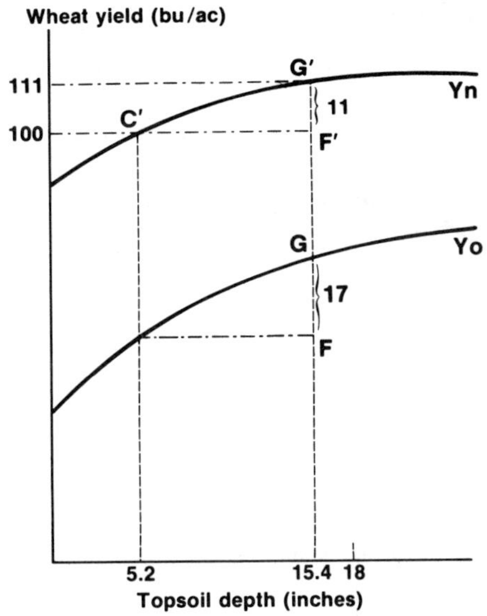

FIGURE 6 Residual yield damage with land-substituting technical change.

Induced Technology

The discussion thus far has assumed that any technical advance would occur independently of farmer decisions about conservation and the resulting rate of erosion. There is a possibility, however, of induced technical progress. Concern over the rate of erosion might encourage research and development that results in yield enhancing technical advances.

Induced technology alters the damage assessment procedure. Recall that the relevant comparison (dynamic basis) for damage assessment with exogenous technology is yield with technology on conserved soil versus yield with the same exogenous technology on eroded soil. Yield damage is measured along the single technology-augmented yield function. Only one yield function is needed because the same level of technology would apply independently of erosion scenario. But if all technical advance is induced, two yield functions are needed to reflect the different levels of technology in the with conservation and without conservation scenarios.

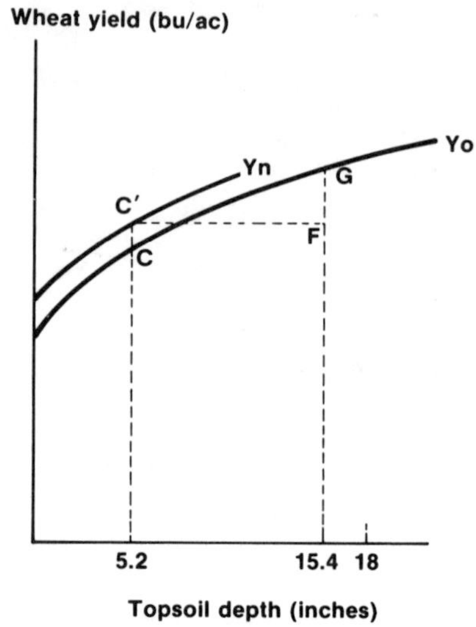

FIGURE 7 Residual yield damage with induced technology.

Damage assessment should be based on yield with conservation and unchanged technology versus yield with erosion and induced technology. Because two yield functions reflecting different levels of technology are involved, induced technology always offsets some erosion damage.

With induced technology, as illustrated by the shift from Yo to Yn in Figure 7, yield damage is the difference between yield at G, conserved topsoil and unchanged technology, versus yield at C', eroded topsoil and induced technology.[3] In the absence of technology, yield damage would have been the difference in yield between G and C in Figure 7. But technology induced by concern over erosion boosts yield from C to C' at the eroded topsoil depth, offseting some erosion damage.

Exogenous and Induced Technology

To exhaust all the possibilities, consider the case with both exogenous and induced technology. Exogenous technology shifts the yield function from Yo to Y_1 in Figure 8. This technical advance would occur even in the absence of heavy erosion, so Y_1 is the correct curve for measuring yield at the conserved topsoil depth. With

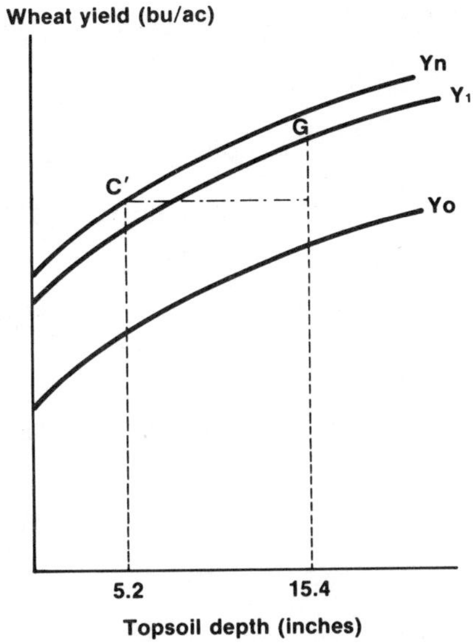

FIGURE 8 Residual yield damage with exogenous and induced technology.

heavy erosion over time, topsoil may be eroded to, say 5.2 inches. Figure 8 incorporates an additional shift in the yield function to Yn from technology induced by concern over heavy erosion. Yield with eroded topsoil is thus measured along curve Yn. The correct damage measure with induced and exogenous technology combined would be yield with conserved soil and exogenous technology at G versus yield with eroded soil and induced technology at C'.

NATURE OF PAST TECHNICAL PROGRESS: EMPIRICAL EVIDENCE FOR WINTER WHEAT YIELDS IN THE PALOUSE REGION

The discussion in the preceding section established the importance of projecting technology trends as well as erosion rates to assess erosion damage accurately. While there is no foolproof method for projecting whether future advances in agricultural technology are likely to be land-neutral, -complementary, or -substituting, and exogenous or induced, a logical first step is to examine how recent technical advances have influenced yields.

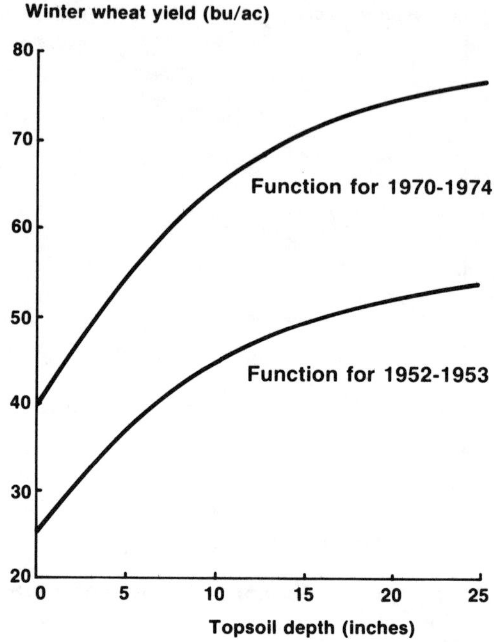

FIGURE 9 Comparison of winter wheat yield-topsoil depth relationships from the 1950s and the 1970s, eastern Whitman County, Washington.
SOURCE: Young et al., 1985.

Young et al. (1985) conducted a statistical evaluation of the impact of technical progress on winter wheat yields in the eastern Palouse region of southeastern Washington between the 1950s and the 1970s. This 2-decade interval witnessed several notable advances in wheat production technology in the region, including introduction of higher yielding semidwarf varieties, greater use of commercial nitrogen fertilizer, more effective chemical weed control, and improved tillage techniques.

Figure 9 summarizes the statistical functions that describe the response of winter wheat yields to topsoil depth in the eastern Palouse during the early 1950s and the early 1970s. The 1950s function was derived by Young et al. (1985) from statistical relationships estimated by Pawson et al. (1961) using over 800 observations from farmers' fields collected during 1952 and 1953. The 1970s function was estimated by Taylor (1982) from 89 observations, also from farmers' fields, collected by Wetter (1977) in the same region during 1970-1974.

Equations 1 and 2 describe the statistical relationships underlying the response functions in Figure 9:

$$\text{1952-1953 function } Y = 24.96 + 31.64(1-0.90^{D}) \quad (1)$$

$$\text{1970-1974 function } Y = \underset{(3.40)}{38.92} + \underset{(4.79)}{40.50}(1-0.90^{D}), \quad R^2 = 0.45, \quad (2)$$

where Y and D represent winter wheat yield in bushels per acre and topsoil depth in inches, respectively.

R^2 is the proportion of wheat yield variation in the data set explained by the 1970-1974 curve in Figure 9. The figures in parentheses under the coefficients of Equation 2 are standard errors. Pawson et al. did not report these for their equation but Young et al. hypothesized standard errors of equal magnitude for the two functions to test statistically the nature of the technology shift between them. The results of this test rejected at the 10 percent significance level the hypothesis of a land-neutral or land-substituting technology shift and supported the alternative of land-complementary technical change over the 2 decades.

Indeed, point estimates of yield projections from the two functions reveal that technology over the 2 decades boosted average wheat yields by 22.9 bushels/acre on a deep 30-inch topsoil, but only by 14.4 bushels/acre on subsoil (0 inches topsoil). This represents a 59 percent greater yield increase due to technology on the deeper topsoil. Kaiser (1967) provides similar evidence of greater wheat yield growth in the Palouse on deeper topsoils based on unpublished data from the 1950s and 1960s.

The land-complementary technical shift evidenced in Figure 9 is consistent with agronomic principles. Among other properties, the new semidwarf wheat cultivars have greater genetic potential for converting moisture and nutrients to harvestable grain. However, these cultivars are likely to come closest to achieving their higher genetic yield potential in an uneroded soil environment where moisture, nutrients, and rooting zone are usually more suitable for crop growth. Furthermore, the higher yield potential with new cultivars or improved cultural practices will be restricted at the outset if soil structure problems on clay subsoils exposed by erosion impede germination and establishment of the stand.

Support for the view that future technical progress in the region also is likely to be land-complementary comes from one other important group--the farmers in the region. A survey of 272 Palouse farmers in 1980 revealed that on average the farmers expected wheat yield growth over the next 50 years to be three times higher on typical hillslopes than on hilltops, which are more eroded and have much shallower topsoils (STEEP Project, 1980).

The empirical yield response functions in Figure 9 should also permit conclusions concerning residual yield damage, as described in the previous section. Farmers, whose fields were included in the sample used to estimate the 1970's function, have presumably adjusted inputs in a profit-maximizing manner in response to erosion and other changes over the 2 decades. Consequently, the 1970's response function should represent the conceptual restored yield function described in Figure 3, as required to measure residual damage.

Finally, it seems that most, if not all, of the technical progress in wheat production that occurred in the Palouse between the early 1950s and early 1970s was exogenous, not induced specifically by concern over erosion. Topsoil depletion was apparently much less important than other factors in determining agricultural technology in this period. For example, concerns about lodging and disease resistance were major factors in development of the short-strawed semidwarf wheat varieties. The development of inexpensive procedures for producing inorganic nitrogen fertilizers and effective herbicides grew out of pervasive exogenous breakthroughs in chemical technology during and after World War II.

Furthermore, the development of improved crop cultivars is not an advance applicable only to eroded fields. In fact, as mentioned, uneroded soils are a more suitable environment. Fertilizer application is also widely practiced and is not used solely on eroded sites. Because of impairment in soil structure or moisture limitations with erosion on some soils, fertilizer technology would boost yields more on uneroded sites with these soils. It is not likely that technologies with a greater payoff potentially on uneroded sites would be induced by concern over erosion. It is more likely that these technologies are the result of a desire to enhance agricultural productivity in general and therefore must be considered exogenous.

A final reason for believing that much of the technical progress in yields has been exogenous is the difficulty in applying embodied technologies differentially to eroded and uneroded parts of a field. It would not be practical for farmers to plant a special hybrid variety on eroded sites in a field and a standard variety in the balance of the field. Similarly, it is often not feasible to vary fertilizer application on eroded sites within a field. Given the difficulty in treating eroded parts of fields differently with these techniques and since these technologies increase yield more on deeper, uneroded soils in many cases, it seems fair to conclude that much of the yield-enhancing technology in agriculture during this century has been exogenous rather than induced by erosion.

Hayami and Ruttan (1971), who are noted for their studies of technical progress in agriculture, reach similar conclusions for U.S. agriculture in general. They ascribe much of the yield progress in agriculture beginning with the 1930s to higher yielding varieties such as hybrid corn and to the development of commercial nitrogen fertilizers. They present evidence to support the claim that the development and improvement of these yield-enhancing technologies was stimulated by general fertility limitations of the land as reflected in the stagnant yield trend from the 1870s to the 1920s. If the concern that motivated scientific and commercial interest in developing these technologies was concern about general fertility limitations in the agricultural land base, these technical advances must be considered exogenous, not induced by concern over soil erosion.

The consequences for erosion damage assessment are significant. Much of the yield-enhancing technical progress in agriculture has <u>not</u> been the type that offsets erosion damage. As shown earlier in this paper, exogenous, land-complementary technical progress increases erosion damage. Thus, technology in the Palouse, and perhaps in the rest of the nation, rather than mitigating the problem of erosion and yield damage, has actually intensified it.

Appendix B develops and examines an empirical Palouse wheat yield projection function incorporating the dual influences of erosion and technology consistent with the data and principles reviewed in this section.

COMPUTERIZED DAMAGE ASSESSMENT MODEL

A computerized model has been developed for assessing the economic cost of erosion damage incorporating the principles outlined thus far (see Appendix C). The earlier discussion of physical yield damage used an erosion period of many years to illustrate damage graphically. The economic erosion damage model is programmed to assess the incremental damage from one more year of erosion, where the consequences of that erosion are measured over a future damage horizon.

The model contains two main components--a time-driven erosion productivity simulator and an economic assessment module. The erosion productivity simulator models the physical relationship between erosion and soil properties and then models the impact of those changed soil properties on crop yields. The current version of the erosion productivity simulator (for analysis in the Palouse) uses topsoil depth (depth of the mollic epipedon[4]) as a proxy for soil properties such as organic matter content and bulk density that are affected by erosion and are correlated with topsoil depth. In addition to modeling the negative impact of erosion on crop yields, this simulator also projects the positive impact of technical progress on yields.

The economic assessment module evaluates the long-run and short-run economics of erosion control. Long-run economics encompass the cost in the future of damage from current erosion. One such cost is the present value of lost future income over a relevant damage horizon from reduced yields due to erosion in the current year. In the past, a damage horizon of 75 years has been used for evaluating the future consequences on yield and income of current-year erosion.

This time horizon is long enough to incorporate the management periods of current operators, their children, and their grandchildren. With family farms, it is reasonable to assume that an operator would be concerned about those future consequences of his management decisions. Also, with a 4 percent real private rate of discount, a 75-year time horizon captures 95 percent of the present value of erosion damage into perpetuity. Another component of long-run damage is the present value of the cost of any soil-substituting inputs, such as fertilizer, that are increased in the future to offset the effect of current erosion on future crop yields. These two cost streams constitute the residual and reparable erosion damage components, respectively.

A short-run economic evaluation can be included if the erosion damage assessment employs the best conserva- tion alternative as the basis for comparison. It compares the current income of the erosive practice with that of the conservation practice. The value of any yield differential with the conservation practice (due to a tillage yield penalty, for example) is captured here as well as any cost difference between the practices.

The damage model was used to make an empirical estimate of the cost of erosion damage for wheat production with conventional tillage in the Palouse. Annual soil loss with this practice on typical slopes averages 10.4 tons/acre. The damage estimate presented here is based on a comparison of "with versus without" erosion (as explained in the first section of this paper) and measures the present value of the lost income over 75 years from reduced future yields due to 1 year of erosion with conventional tillage in wheat.

The price of wheat was $3.60/bushel and the initial topsoil depth was assumed to be 10 inches, a moderately to severely eroded soil in the Palouse. The empirically estimated wheat yield projection function (see Appendix B), incorporating exogenous land-complementary technology observed in the Palouse, was used in the damage model.

The cost of erosion damage under the assumed conditions is $12.78 per acre. This is the present value of the lost income from 1 year of erosion with conventional tillage in wheat. The cost of erosion damage would be less with deeper topsoil and greater with shallower topsoil.

USE OF NRI DATA FOR REGIONAL AND NATIONAL EROSION DAMAGE ASSESSMENT

Applications to date of the economic damage model described in the previous section and in Appendix C have been exclusively at the farm level (Walker, 1982; Walker and Young, 1986). These have generated results on the optimal point in time for farmers to adopt specified conservation practices. These results could be useful in SCS conservation education programs with farmers when private onsite benefits from avoiding erosion damage justify immediate adoption of conservation practices.

Results have also been generated for policymakers on the required conservation subsidy (or erosion penalty) necessary to encourage farmers to adopt conservation practices when social criteria warrant adoption earlier than would be optimal from a strictly private evaluation of onsite effects.

While development of region-specific conservation education and incentive programs is likely to remain a major use of these damage assessment techniques, the general concepts also have relevance for national or regional assessments of onsite productivity damage from erosion. Walker (1983) has proposed a net benefit function for use in selecting target areas for conservation emphasis. Such areas could be identified based on the economic cost of erosion damage and the potential for avoiding that damage with appropriate conservation practices that are available for the area. This section evaluates the possible use of information from the 1982 NRI for national economic damage assessment and identifies other data requirements and possible sources.

NRI DATA RELEVANT FOR DAMAGE ASSESSMENT

The 1982 NRI, like its predecessors, focused primarily on describing current use, annual erosion rates, conservation treatment needs, and cropland conversion potential of all private land in the United States.[5] Given its primary emphasis on these land characteristics, the NRI would not be expected to contain much of the detailed information on crop productivity relationships required to assess economic erosion damage. As described earlier, measuring erosion damage requires detailed information on crop yields by erosion status (e.g., remaining topsoil depth) for different tillage systems. Furthermore, to make inferences about the interaction of erosion and technology over time, time-series observations are needed on site-specific crop yields. However, no crop yield information was collected for the 1982 (or earlier) NRI sample sites.

The NRI data base does contain four other information components required for national erosion damage assessment:

- Estimates of erosion rates by region and site for prevailing management systems. Both the site characteristics and management practices are described in considerable detail. These include explicit values for all the variables of the Universal Soil Loss Equation (USLE). Management data collected include land use, irrigation status, cropping history for the past 3 years, conservation practices used, and value of the USLE crop management factor.

- Degree of past erosion. Information was collected on "degree of erosion," whether the site was "nonarable because of past erosion," "soil loss tolerance limit" or T value, and "land capability subclass." Possibly this information, plus supplementary data from local soil surveys, could be used to estimate current topsoil depths for different soils in a region.
- Information on existing and needed conservation treatments. As noted earlier in this paper, measurement of damage averted by conservation requires identification of the optimal (most cost-effective) conservation farming system for a particular area. Data on existing and needed conservation treatment for each site might help identify the optimal conservation system for different areas.
- Topographic features and cropping patterns. The very detailed information on distribution of cropland by topographic features and cropping patterns throughout the nation would be useful for aggregating the cost of erosion damage.

ADDITIONAL DATA NEEDS FOR DAMAGE ASSESSMENT

Along with the technical information available from the NRI data base, regional or national erosion damage assessments will require assembling data (or assumptions) on:

- Static technology yield relationships
  --Yield penalties for alternative management systems
  --Crop yield impacts of erosion within a given technological era
- Technical progress relationships
  --Whether technical progress for various regions and crops is induced or exogenous with respect to erosion
  --Whether technical progress for various regions and crops is land-substituting, -neutral, or -complementary
  --Projected rates of technical progress, for various regions and crops

- Cost and returns information
  --Crop prices through time
  --Current production costs for different management systems
  --Changes in production costs as optimal input adjustments are made in response to reparable erosion damage
- Present value analysis parameters
  --Discount rates
  --Planning horizon lengths

The yield relationships can either be estimated empirically (as exemplified by the Palouse wheat yield functions depicted in Figure 9) or synthesized using general simulation models such as the Erosion Productivity Impact Calculator (EPIC) model developed by Agricultural Research Service scientists at Temple, Texas (Williams et al., 1983).

Yield relationships must be derived by uniform procedures for all crops and regions to obtain consistent national erosion damage estimates. This means that synthetic yield projections as generated by the Yield-Soil Loss Simulator [used for the 1980 Resource Conservation Act (RCA) Appraisal] or the EPIC model (used for the 1985 RCA Appraisal) will probably be necessary. Appropriate data sets for estimating empirical topsoil depth-yield relationships are unlikely to be available or affordable for all major crops and production regions. Where appropriate data are available, these relationships should be estimated to validate and calibrate the general simulation models.

Looking ahead, incorporating soil depth and yield measurements into future NRIs would provide a consistent national data base for estimating yield relationships. Although most previous analyses have used topsoil depth alone as a proxy for the set of soil properties altered by erosion (Young, 1984), other soil properties like organic matter content may also be measured if necessary to make accurate yield projections.

As indicated earlier, forecasts of the rates and nature of future technical progress for various crops and regions are necessarily subjective, but evaluation of historical trends as exemplified by this analysis of winter wheat yields in the Palouse can provide some guidance. As noted in a recent review by Young (1984) of crop yield projection models employed in 15 long-run soil conservation benefit evaluations, there has been little or no investigation outside the work reported here from the Palouse on technical progress-topsoil depth interactions or on whether technology was induced by erosion or exogenous. Given its importance for damage assessment, high priority should be given to searching for data sets to examine technical progress patterns for other major crops and production regions.

Assessment of erosion damage in the 1980 RCA appraisal incorporated crop- and region-specific rates of technical progress (USDA, 1981), but assumed uniformly multiplicative technology--a form of land-complementary technology--throughout, as shown by Young (1984) based on documentation in Benbrook (1980). Also, the 1980 RCA appraisal implicitly assumed exogenous technology, as evidenced by the use of a single technology-augmented yield function for measuring erosion damage with a crop in a production region. Although the crop- and region-specific technology rates were modeled with some detail, little or no judgment was made on whether technical progress was (1) exogenous or induced or (2) land-complementary, -substituting or -neutral across crops and regions.

If technology differs from this assumed uniform pattern for some crops and regions, substantial bias in damage projections could result. In future RCA appraisals, it would be desirable to elicit forecasts of technical progress patterns from agricultural scientists who are familiar with past technical advances in crop yields for major production regions and are knowledgeable about likely developments in the foreseeable future.

Information on base-period crop prices can be obtained from statistical reporting services in the states. Supply and demand projections would be required to model endogenous changes in crop prices through time in response to erosion impacts. Base-period crop production costs for different management systems can be obtained from budgets prepared by extension economists and others. The EPIC model contains a submodel option that computes required fertilizer adjustments to compensate for fertility losses due to erosion. This might provide a basis for estimating reparable erosion damage.

Discount rates and planning horizons should be elicited from the appropriate decision-making clientele--farmers for studies providing private managerial recommendations and policymakers for social evaluations.

SUMMARY AND CONCLUSIONS

Four important concepts were developed and presented for correctly assessing erosion damage to crop productivity: (1) use a "with versus without" comparison in measuring erosion damage to yields (yield with conservation versus without, or yield with erosion versus without); (2) avoid confounding conservation tillage yield penalty and erosion damage; (3) distinguish between reparable and residual yield damage, and include both components in the cost of erosion damage; and (4) project the separate effects of erosion and technology to avoid errors in erosion damage assessment caused by confounding erosion and technical progress. A computerized erosion damage model was described that incorporates these concepts.

Ignoring technical progress in erosion damage assessment can lead to serious bias. Concluding categorically that technology offsets erosion damage because of a positive yield trend over time is a naive view based on an assumption that yield enhancement is due to technological change induced by concerns over erosion. Exogenous technology can mitigate erosion damage only if it is land-substituting.

Economists at Resources for the Future, citing Hayami and Ruttan (1971), have concluded that little agricultural technology has been induced nationwide by concern about the effect of erosion on productivity (Crosson and Stout, 1983). If yield-enhancing technical progress in major producing areas has indeed been exogenous and continues to be, then technogical progress has not erased erosion damage and is not likely to do so in the future. Thus, technology must not be seen generally as a substitute for soil conservation. In fact, if the exogenous land-complementary technology observed over the past 30 years in the Palouse continues and is typical of the situation across the country, technology--by boosting yields more on deeper topsoils--will increase the payoff from soil conservation.

Another way of viewing this conclusion is that the cost of erosion damage is likely to increase in the future. Even without technical advance, the nonlinear yield response curve gets steeper as cumulative erosion reduces topsoil depth. If land-complementary technology continues, the yield curve will become steeper everywhere. A given amount of erosion reduces yield more if the yield response function is steeper, leading to a greater cost of erosion damage. Thus, soil conservation may become even more important in the future.

The evidence on the exogenous, land-complementary technical progress affecting winter wheat yields in the Palouse confirms that technology has complemented, not substituted for, soil conservation. A prudent strategy for ensuring future productivity includes continued support both for basic research to promote future technical progress and for vigorous soil conservation programs. Improved soil conservation enhances the payoff on research and development and vice versa.

With the prospect of increased erosion damage costs in the future, an accurate assessment of erosion damage and using that information to target conservation efforts become all the more important. It may be sufficiently important to justify including additional data items in future NRI surveys to measure productivity impacts from erosion and to infer yield-enhancing technology trends for correct damage assessment.

NOTES

1. These concepts for measuring erosion damage were originally presented in a paper by Walker (1983). The implications of induced technology for erosion damage, which were not discussed in that paper, are developed fully in this paper.
2. For a general mathematical proof of these conclusions that does not rely on specific graphical examples, see Young et al. (1985).
3. With induced technology, the generally improper comparison for assessing damage (yield before erosion versus yield after) coincides with the proper comparison (yield with erosion versus yield without). With exogenous technology, the two measures do not coincide. To assess damage correctly in all cases, use the "with versus without" comparison.
4. The mollic epipedon refers to the darkened upper layer of soil material with high concentration of organic matter. This layer includes the A horizon and may include a transitional B horizon.
5. The 1982 NRI is described in various USDA and National Research Council publications (National Research Council, 1982; USDA, n.d., 1983, 1984). Burns and Dunford (1985) also provide a concise description of the content and procedures of the 1982 NRI, including a copy of the questionnaire completed for each sample point.

REFERENCES

Benbrook, C. 1980. Review of the yield-soil loss simulator. Council on Environmental Quality, Washington, D.C. Memorandum.

Burns, S. H., and R. W. Dunford. 1985. An Overview of the 1982 National Resources Inventory. A. E. Series 85-2. Department of Agricultural Economics. Pullman: Washington State University.

Crosson, P. R., with A. T. Stout. 1983. Productivity Effects of Cropland Erosion in the United States. Baltimore, Md.: Johns Hopkins University Press.

Hayami, Y., and V. Ruttan. 1971. Agricultural Development: An International Perspective. Baltimore, Md.: Johns Hopkins University Press.

Kaiser, V. G. 1967. Soil erosion and wheat yields in Whitman County, Washington. Northwest Science 41(2):86-91.

National Research Council. 1982. Review of the National Resources Inventory Methods and Procedures. A report prepared by the Task Force on the National Resource Inventory, Board on Agriculture and Renewable Resources. Washington, D.C.: National Academy Press.

Pawson, W. W., O. L. Brough, Jr., J. P. Swanson, and G. M. Horner. 1961. Economics of Cropping Systems and Soil Conservation in the Palouse. Agricultural Experimental Station Bulletin 2. Pullman: Washington State University.

STEEP Project. 1980. A Survey on Crop Production in the Palouse. Washington State University and University of Idaho, Departments of Agricultural Economics and Rural Sociology. Unpublished.

Taylor, D. B. 1982. Evaluating the Long Run Impacts of Soil Erosion on Crop Yields and Net Farm Income in the Palouse Annual Cropping Region of the Pacific Northwest. Ph.D. dissertation. Washington State University, Pullman.

USDA (U.S. Department of Agriculture). 1981. Soil and Water and Related Resources in the United States: Analysis of Resource Trends. 1980 RCA Appraisal, Part II. Washington, D.C.: U.S. Government Printing Office.

USDA. 1983. Release of 1982 National Resources Inventory Preliminary Data: Executive Summary. Washington, D.C.: U.S. Government Printing Office.

USDA. 1984. National Resources Inventory: A Guide for Users of 1982 NRI Data Files. (Draft.) Washington, D.C.: U.S. Government Printing Office.

USDA. n.d. SCS. Inventory and Monitoring. National Resources Inventory: Sampling Design. Washington, D.C.: U.S. Government Printing Office.

Walker, D. J. 1982. A damage function to evaluate erosion control economics. Am. J. Ag. Econ. 64(4):690-698.

Walker, D. J. 1983. Targeting soil conservation with a net benefit function incorporating erosion damage and technology. Paper presented at Symposium in Honor of John F. Timmons, Iowa State University, October 28, 1983. A. E. Series 396, Department of Agricultural Economics, University of Idaho, Moscow.

Walker, D. J., and D. L. Young. 1986. The effect of technical progress on erosion damage and economic incentives for soil conservation. Land Economics. In press.

Wetter, F. 1977. The Influence of Topsoil Depth on Yields. Tech. Note AGRO-10, and unpublished data underlying report. Soil Conservation Service, Colfax, Wash.

Williams, J. R., K. G. Renard, and P. T. Dyke. 1983. EPIC: A new method for assessing erosion's effect on soil productivity. J. Soil Water Conserv. 38(5):381-383.

Young, D. L. 1984. Modeling agricultural productivity impacts of soil erosion and future technology. Pp. 60-85 in Future Agricultural Technology and Resource Conservation, B. B. English, J. A. Maetzold, B. R. Holding, and E. O. Heady, eds. Ames: Iowa State University Press.

Young, D. L., D. B. Taylor, and R. I. Papendick. 1985. Separating erosion and technology impacts on winter wheat yields in the Palouse: A statistical approach. In Erosion and Soil Productivity. St. Joseph, Mich.: American Society of Agricultural Engineers.

APPENDIX A: EROSION-COMPENSATING TECHNOLOGY

A special case of land-substituting technology enhances the capacity to repair erosion damage. This special case involves a new or improved input that is applied in increasing quantities on eroded soil. Because this type of technology specifically remedies a deficiency in a soil attribute caused by erosion, it is called erosion-compensating technology. This technology, by its nature, increases reparable damage. But it reduces residual damage by more, so that overall yield damage is reduced, as will be illustrated.

Because this technology reduces residual yield damage, it is considered to be a special case of land-substituting technology. In the general case of land-substituting technology, the new or improved input is applied at a uniform rate for all topsoil depths. The entire yield function shifts upward but in a fashion that reduces the slope of the restored-yield curve. Because technology interacts with topsoil depth to boost yields more on shallow eroded soils (even though the application is uniform across topsoil depths), this general case was classified as land-substituting. In the pure erosion-compensating special case, none of the new or improved input would be used on deep soils but increasing quantities would be used on eroded soils.*

In this pure special case, technology rotates the restored-yield function through point A (deep topsoil) in Figure A-1. Suppose initially that this technology is exogenous, not induced by erosion. The upper terminus of the relevant constant-input yield curve shifts from G to G'. The relevant curve is the one associated with the conserved topsoil depth, D, which is the basis for

*It is also possible that some of the new or improved input might be used on the deepest soil with application increasing at shallower depths. This technology would shift the yield function upward, as in Figure 6 in the text, but unlike the pure general case of land-substituting technology, this mixed case would involve increased application at shallower depths. The analysis of residual and reparable damage would proceed exactly as in the pure special case of erosion-compensating technology presented here.

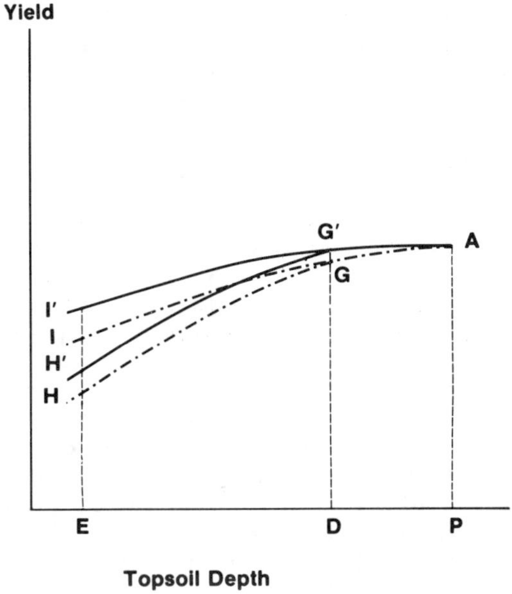

FIGURE A-1 Erosion-compensating technical progress.

comparison in estimating erosion damage. Potential yield on conserved soil shifts with improved technology from G to G' because of the new or improved input associated with the technical advance. That same input level is applied for all topsoil depths along the constant-input yield curve, shifting it from GH to G'H'. The comparison of the shift from H to H' at the eroded topsoil depth E to the shift from G to G' depends on the change in the marginal product of that new or improved input with topsoil depth. The change in marginal product with respect to topsoil depth is given by:

$$\partial^2 y/(\partial x_s \partial x_i) < 0, \quad \text{(A-1)}$$

where y equals $f(x_1,\ldots,x_s)$, crop yield is a function of a vector of inputs; x_i equals the input associated with the technical advance; and x_s equals topsoil depth.

The marginal product of the new or improved input increases with decreased topsoil depth because the input is a substitute for topsoil depth. The input is applied in the same quantity at eroded and conserved topsoil depths along the constant-input yield curve, yet because marginal product increases with shallower topsoil, the constant-input yield curve shifts upward slightly more at the eroded depth than at the conserved depth, HH' > GG'. As a result, the overall yield damage decreases slightly with exogenous erosion-compensating technology, G' - H' < G - H.

The decrease in overall yield damage is the result of two other changes--an increase in reparable damage but a decrease in residual damage of a larger magnitude. The increase in reparable damage, Δ_B = II' - HH' > 0, and the decrease in residual damage, Δ_D = -II' + GG' < 0, are illustrated in Figure A-1.

Reparable damage increases with erosion-compensating technology because the upper curve shifts more than the lower curve, II' > HH'. None of the new or improved input associated with erosion-compensating technology is used at I or H (before technology). With technical advance, the amount of the input used at I' on the restored-yield curve is greater than the amount used at H' on the constant-input yield curve. Because of the greater application of the new or improved input, the shift II' exceeds the shift HH' and reparable damage increases with erosion-compensating technology, I' - H' > I - H.

Residual damage decreases with erosion-compensating technology because the restored-yield curve shifts upward more at shallow topsoil depths, II' > GG'. More of the erosion-compensating input is used for the shift II' and its marginal product is higher than at GG'. Therefore, residual damage decreases, G' - I' < G - I.

Comparing absolute values shows that the decrease in residual damage is greater than the increase in reparable damage:

$$|\Delta_B| = |II' - HH'| = II' - HH'$$
$$|\Delta_D| = |-II' + GG'| = II' - GG'$$
$$|\Delta_D| > |\Delta_B| \text{ because } HH' > GG'.$$

To recapitulate, exogenous erosion-compensating technology increases reparable damage, but reduces residual damage more; as a result, overall yield damage decreases.

If the shift in yield functions illustrated in Figure A-1 were due to induced technology, residual damage would decrease even more. Using the appropriate "with versus without" conservation comparison, residual yield damage would be the difference between yield with conservation versus yield with erosion and induced technology, G - I'. With induced erosion-compensating technology, residual damage is less than with exogenous technology, G - I' < G' - I'. Reparable damage is the same as with exogenous technology because reparable damage is measured at the eroded topsoil depth, where the yield function shift from technology is the same whether it is induced by erosion or exogenous. Because induced technology reduces residual damage more than exogenous technology does while the effect on reparable damage is the same, induced erosion-compensating tech- nology reduces overall yield damage more than exogenous technology does, G - H' < G' - H'. Erosion-compensating technology may often be induced rather than exogenous because it specifically remedies a soil property altered by erosion.

Even though reparable damage increases, the cost of erosion damage decreases for two reasons. First, an outright decrease in residual yield damage equal to $|\Delta_D| - |\Delta_B|$ reduces the cost of erosion damage by the value of that yield. Second, residual damage is replaced by reparable damage in the amount $|\Delta_B|$. This, too, reduces the cost of erosion damage because the cost of the remedy must be less than the value of the residual yield damage that is restored or repaired.

APPENDIX B: WHEAT YIELD PROJECTION FUNCTION

This appendix discusses a wheat yield projection function for the Palouse that exhibits the historical pattern of interaction between technology and topsoil depth described in the text. This function was introduced initially by Papendick et al. (1985), but it is developed more fully here. The projection function is then incorporated into a computerized erosion damage assessment model.

Algebraically, response functions such as those illustrated in Figure 9 in the text can be expressed as:

$$Y_D = a + b(1 - R^D) \text{ for the base period} \qquad \text{(B-1)}$$

$$Y_D = A + B(1 - R^D) \text{ for T years later,} \qquad \text{(B-2)}$$

where Y_D equals wheat yield in bushels per acre; (a, b, A, B, R) are estimated parameters with (a, A) $\geq$ 0; (b, B) > 0, and $0 < R < 1$; and D equals topsoil depth in inches. Taking advantage of the common functional form of (B-1) and (B-2), the two equations can be combined into a single yield projection function by including a time variable, t:

$$Y_t = (a + a't) + (b + b't)(1 - R^D), \qquad \text{(B-3)}$$

where Y_t is projected yield in year t, t = 0 represents the base period characterized by Equation B-1, (a, b, R, and D) are as defined above, and

$$a' = (A - a)/T, \text{ and} \qquad \text{(B-4)}$$

$$b' = (B - b)/T. \qquad \text{(B-5)}$$

Equations B-4 and B-5 preserve the exact historical pattern and rate of technical progress on the "intercept" and "slope" coefficients of the response functions illustrated in Figure 9 in the text. For simplicity, a' and b' incorporate the average annual rate of adjustment in the two coefficients that was observed during the T years separating the two functions. Both a' and b' are positive, assuming technical progress. Taking the time derivatives of Equation B-3 as topsoil depth approaches 0 and infinity, a' and (a' + b'), respectively, can be interpreted as the annual rates of change in wheat yields due to technology on subsoils and very deep topsoils. Topsoil depth, D in Equation B-3, can also be expressed as a function of time:

$$D = D_o - A_s t, \qquad \text{(B-6)}$$

where D_o is the original topsoil depth in the base year, when $t = 0$, and A_s is the erosion rate expressed in inches per year. Substituting Equation B-6 into Equation B-3 provides the final yield projection function:

$$Y_t = (a + a't) + (b + b't)[1 - R^{(D_o - A_s t)}]. \qquad \text{(B-7)}$$

Equation B-7 describes the combined impact of technical progress and erosion on wheat yields given the pattern of recent technical progress for winter wheat in the Palouse.

PROPERTIES OF THE PROJECTION FUNCTION

Before using a mathematical projection function such as B-7 for long-term simulations, it is important to examine the plausibility of its mathematical behavior.

Extremum and Slope Conditions

Will yields continue growing for a period, reach a peak, and then begin declining eventually as a result of the joint influence of erosion and technical progress incorporated in Equation B-7? To answer this question, differential calculus is used to examine the annual growth (or decay) rate for yield:

$$dy_t/dt$$

$$= d[(a + a't) + (b + b't)(1 - R^{(D_o - A_s t)})]/dt$$

$$= d[a + a't + b - bR^{(D_o - A_s t)} + b't - b'tR^{(D_o - A_s t)}]/dt$$

$$= a' + bR^{(D_o - A_s t)}(\ln R)A_s + b' - b'[R^{(D_o - A_s t)}$$

$$+ R^{(D_o - A_s t)}(\ln R)\ (-A_s)t]$$

$$= \underbrace{(a'+ b')}_{>0} + \underbrace{\overbrace{R^{(D_o - A_s t)}}^{>0}[\overbrace{b(\ln R)A_s}^{<0} \overbrace{- b'}^{<0} + \overbrace{b'(\ln R)A_s t}^{<0}]}_{<0}. \quad \text{(B-8)}$$

Given the presence of technical progress, continuing soil erosion, and the restrictions on parameters imposed at the outset, the definitive sign determinations noted in Equation B-8 can be made. Note that $(\ln R) < 0$ because $0 < R < 1$.

Clearly, the yield trajectory can either be rising or declining depending upon the value of t and the parameters. Whenever $(a' + b')$ in Equation B-8 is greater in absolute value than the following negative term, yields will be rising over time. When this inequality is reversed, yield decline will occur over time.

The point in time at which yields peak out (if such a point exists) might be identified by evaluating the first-order conditions for a local extremum.

$$dy_t/dt = (a' + b')$$

$$+ R^{(D_o - A_s t)}[b(\ln R)A_s - b' + b'(\ln R)A_s t] \overset{\text{set}}{=} 0, \quad \text{(B-9)}$$

which implies

$$R^{(D_o - A_s t)} = (a' + b')/[-b(\ln R)A_s + b' - b'(\ln R)A_s t],$$

which implies

$$(D_o - A_s t)\ln R = \ln(a' + b')$$

$$- \ln[-b(\ln R)A_s + b' - b'(\ln R)A_s t]. \quad \text{(B-10)}$$

Although Equation B-10 implicitly defines a function of t, the expression cannot be explicitly solved for t by algebraic procedures. However, unreported numerical analyses of the yield-projection function with parameters typical of those for winter wheat grown on eastern Palouse sites showed the yield trajectory to peak out. For different response parameters, technical progress rates, or erosion rates, the results could be quite different.

Concavity

It is also of interest to evaluate the concavity of the yield projection function. Taking the second time derivative of Equation B-7 yields:

$$d^2y_t/dt^2$$

$$= d\{(a' + b') + R^{(D_o - A_s t)}[b(\ln R)A_s - b' + b'(\ln R)A_s t]\}/dt$$

$$= b(\ln R)A_s R^{(D_o - A_s t)}(-A_s)\ln R - b'R^{(D_o - A_s t)}(-A_s)\ln R$$

$$+ b'(\ln R)A_s[R^{(D_o - A_s t)}] + R^{(D_o - A_s t)}\ln R(-A_s)b'(\ln R)A_s t$$

$$= (\ln R)A_s R^{(D_o - A_s t)}[b(-A_s)\ln R + b' + b' - b'(\ln R)A_s t]$$

$$= \underbrace{(\ln R)A_s R^{(D_o - A_s t)}}_{<0}\{\underbrace{[A_s(\ln R)](-b - b't)}_{>0} + \underbrace{2b'}_{>0}\} < 0. \quad \text{(B-11)}$$

Given the presence of technical progress, continuing erosion, and the restrictions on the parameters imposed at the outset, we can make the definitive sign determinations noted in Equation B-11. Because this second derivative is negative, the yield projection equation proposed in this paper will always generate strictly concave yield trajectories as illustrated in Figure B-1.

It must again be noted that this functional form is based on limited data on one crop, winter wheat, grown in one region, the relatively high-rainfall eastern Palouse

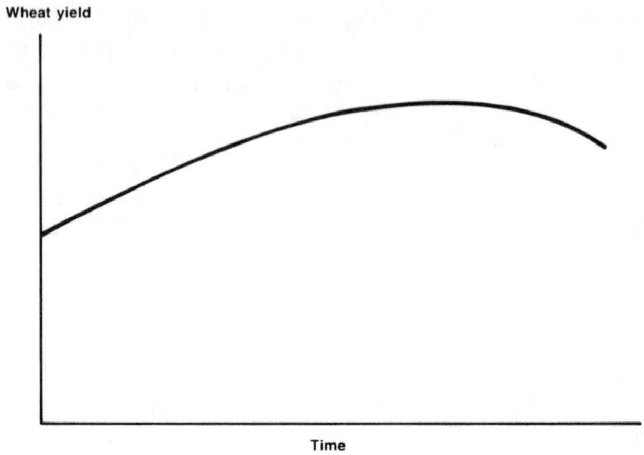

FIGURE B-1 General shape of yield trajectory generated by purposed functional form.

of southeastern Washington. Furthermore, no empirical validation exists for this function in the subsoil (zero topsoil) zone. Under many situations, projected yields could still be rising at the point subsoil is reached. Additional detail on the data sources and statistical analysis techniques underlying the development of this function is provided in Pawson et al. (1961), Wetter (1977), Hoag and Young (1983), and Young et al. (1985).

In general, this examination of Equation B-7 revealed plausible yield trajectories that are concave with respect to time and that can peak and eventually decline if the impacts of erosion outweigh those of technical progress.

EMPIRICAL YIELD PROJECTION MODEL

In specifying the values of a' and b' of Equation B-7, it was decided to use the average wheat yield growth rate for the longer period 1950-1980 instead of that for 1953-1973 (the interval bracketed by the empirical functions in Figure 9 in the text). The period 1953-1973 happened to bracket a period of atypically rapid technical progress in wheat yields in the Palouse. Consequently, use of the shorter period to estimate a' and b' in accordance with Equations B-4 and B-5 was judged likely to overestimate longer-term rates of future technical progress.

The technical progress parameters (a' and b') were calculated as described below to reflect the long-term (1950-1980) effective rate of wheat yield growth in the Palouse. Recall that the yield projection function is:

$$y_t = (a + a't) + (b + b't)(1-R^{(D_o - A_s t)}). \qquad \text{(B-12)}$$

Let a' equal 1.632 b', because this is the ratio exhibited by a' and b' derived by comparing the 1952-1953 response function of Pawson et al. (1961) with the 1970-1974 response function of Taylor (1982) (<u>see</u> Figure 9 and Equations 1 and 2 in the text).

Differential calculus is used to solve algebraically in Equation B-13 for the effective rate of yield growth exhibited by Equation B-12 considering both technical progress and erosion:

$$dy_t/dt \qquad \text{(B-13)}$$

$$= (a' + b') + R^{(D_o - A_s t)}[b(\ln R)A_s - b' + b'(\ln R)A_s t]$$

$$= (2.632b') + R^{(D_o - A_s t)}[b(\ln R)A_s - b' + b'(\ln R)A_s t].$$

The effective yield growth rate in Whitman County has been 0.56 bushels/acre/year over the 1950-1980 period (Homayoun-Mehr, 1982). The rate was faster early in the period and slower later, but this is the long-term average.

By inserting the Pawson et al. and Taylor estimated parameters into the yield projection function and setting $dy_t/dt = 0.56$ for the midpoint year (1965) of the 1950-1980 period, the values of a' and b' can be solved for to yield this long-term historical, and assumed future, rate of yield growth due to technology.

First, set t = 0 at 1952, so t = 13 at 1965. Based on Pawson et al., set D_o = 18 inches as the area- weighted average Palouse topsoil depth in 1952. Following Krauss and Allmaras (1982), 0.059 inches/year is used as the long-term regionwide average soil loss rate in the Palouse.

Then, solve algebraically for b' from:

$$dy_t/dt \overset{set}{=} 0.56$$

$$2.632\ b' + R^{(D_o - A_s t)}[b(\ln R)A_s - b' + b'(\ln R)A_s t] \overset{set}{=} 0.56, \quad \text{(B-14)}$$

which implies

$$b' = [0.56 - R^{(D_o - A_s t)} b(\ln R)A_s]/\{2.632 + R^{(D_o - A_s t)}[-1 + (\ln R)A_s t]\}. \quad \text{(B-15)}$$

Solved at estimated values gives:

$$b' = [0.56 - 0.9^{(18 - (.059)13)} 31.64(\ln 0.9)0.059] \div [2.632 + 0.9^{(18 - (.059)13)}(-1 + (\ln 0.9)(0.059)13)]$$

$$= (0.56 + 0.032)/(2.632 - 0.176)$$

$$= 0.592/2.456 = 0.241 \quad \text{(B-16)}$$

for the 1950-1980 period as opposed to b' = 0.443, when b' is derived solely from the 1952-1972 period using Equation B-5. So the final function, reflecting 1950-1980 long-term technology rates, is:

$$y_t = [24.46 + 1.632(0.241)t] + (31.64 + 0.241t)(1-0.90^{(18 - A_s t)}), \quad \text{(B-17)}$$

where t = -2 at 1950, -1 at 1951, 0 at 1952,..., n at (1952 + n). As an example, projecting eastern Palouse regional average winter wheat yield for 1982:

$$y_{1982} = [24.46 + 0.393(30)] + (31.64 + 0.241(30))(1-0.90^{[18 - 0.059(30)]})$$

$$= 68.34 \text{ bushels/acre.} \quad \text{(B-18)}$$

In fact, 65 to 75 bushels/acre is a widely accepted current average winter wheat yield range for the eastern Palouse.

REFERENCES

Hoag, D. L., and D. L. Young. 1983. Yield-topsoil depth response functions: Linear versus Mitscherlich-Spillman. STEEP Agric. Econ. Working Paper 82-2. Department of Agricultural Economics, Pullman: Washington State University.

Homayoun-Mehr, F. 1982. Analysis of yield trends for Washington pulse and grain crops. Unpublished MA special project paper. Washington State University, Pullman.

Krauss, H., and R. Allmaras. 1982. Technology masks the effects of soil erosion on wheat yields--a case study in Whitman County, Washington. In Determinants of Soil Loss Tolerance, E. L. Schmidt et al., eds. Madison, Wis.: American Society of Agronomy.

Papendick, R. I., D. L. Young, D. K. McCool, and H. A. Krauss. 1985. Regional effects of soil erosion on crop productivity--the Palouse area of the Pacific Northwest. In Soil Erosion and Crop Productivity, R. F. Follet and B. E. Stewart, eds.

Pawson, W. W., O. L. Brough, Jr., J. P. Swanson, and G. M. Horner. 1961. Economics of Cropping Systems and Soil Conservation in the Palouse. Agricultural Experimental Station Bulletin 2. Pullman: Washington State University.

Taylor, D. B. 1982. Evaluating the Long Run Impacts of Soil Erosion on Crop Yields and Net Farm Income in the Palouse Annual Cropping Region of the Pacific Northwest. Ph.D. dissertation. Washington State University, Pullman.

Wetter, F. 1977. The Influence of Topsoil Depth on Yields. Tech. Note AGRO-10, and unpublished data underlying report. Soil Conservation Service, Colfax, Wash.

Young, D. L., D. B. Taylor, and R. I. Papendick. 1985. Separating erosion and technology impacts on winter wheat yields in the Palouse: A statistical approach. In Erosion and Soil Productivity. St. Joseph, Mich: American Society of Agricultural Engineers.

APPENDIX C: COMPUTERIZED EROSION DAMAGE ASSESSMENT FUNCTION

The computerized erosion damage model discussed in this paper is presented mathematically in Equation C-1. The model calculates the present value of the future consequences of choosing the erosive practice for one more year.

$$\delta_t = \qquad \text{(C-1)}$$

$$P.[Y_e(t,D_{t-1}) - Y_c(t,D_{t-1})] - [C_e(t,D_{t-1}) - C_c(t,D_{t-1})]$$

$$- \sum_{i=1}^{T} \{P.[Y_c(t + i,D_{t-1}) - Y_c(t + i,D_t)]$$

$$+ [C_c(t + i,D_t) - C_c(t + i,D_{t-1})]\}/(1 + r)^i,$$

where P equals crop price; D_t equals topsoil depth at the end of year t*; Y_e,Y_c equals crop yield with erosive conventional practice and conservation practice, respectively; C_e,C_c equals production costs of the respective practices (includes variable costs and annualized equipment ownership costs); t equals time variable, which serves as proxy for technology, T equals number of years in damage horizon, and r equals real rate of discount.

This equation is derived in Walker (1982). Time (t) and topsoil depth at the end of period t (D_t) are included as arguments in the yield function to allow projection of yields as a function of the separable effects of erosion and technology. The explicit form of the yield function was developed in Equation B-17. For notational simplicity in Equation C-1, it is assumed that erosion with the conservation practice is negligible so that topsoil depth and yield can be maintained indefinitely with the conservation practice. Walker (1982) also presents a more general formulation of the

---

*An end-of-year convention is used so that topsoil depth and technology at the end of year t-1 are assumed to influence yield and cost in year t.

model where the conservation practice slows but does not eliminate erosion. Time and topsoil depth are also included as arguments in the cost functions to allow inputs and thus costs to vary over time with declining topsoil depth and technology. Price is treated as an exogenous variable in this formulation. In applying the damage model for regional or national erosion damage assessments, it would be desirable to allow for endogenous changes in equilibrium crop prices with cumulative erosion over time.

The terms in Equation C-1 account for the private costs and benefits of choosing the erosive practice one more year and are explained below. The first two groupings of terms capture the effect of tillage choice on current-year income.

$$P.[Y_e(t,D_{t-1}) - Y_c(t,D_{t-1})] \qquad \text{(C-a)}$$

Expression C-a reflects any yield differential between the erosive and conservation practices in the current year. If the erosive practice is higher yielding, this term is positive, representing a benefit in the current year of choosing the conventional practice. If the conservation practice is higher yielding, this term is negative, representing a cost of choosing the conventional practice.

$$-[C_e(t,D_{t-1}) - C_c(t,D_{t-1})] \qquad \text{(C-b)}$$

Expression C-b captures any difference in cost between the two practices in the current year. If the conservation practice saves labor or equipment, this component might be negative--a cost of choosing the erosive practice. If the conservation practice requires more costly chemical weed control, this component might be positive--a benefit from choosing the erosive practice.

The final group of terms captures the impact of tillage choice in the current year on future income.

$$- \sum_{i=1}^{T} \{P.[Y_c(t + i,D_{t-1}) - Y_c(t + i,D_t)]$$

$$+ [C_c(t + i,D_t) - C_c(t + i,D_{t-1})]\}/(1 + r)^i \qquad \text{(C-c)}$$

Expression C-c measures the future cost of erosion damage incurred in the current year or, in the parlance of resource economics, the user cost of exploiting the soil. The first bracketed term is the residual damage from current-year erosion--the present value of lost income in the future from reduced yield due to postponing the adoption of conservation another year. Therefore, yield damage is computed using the conservation yield function. The second bracketed term is the reparable damage due to current-year erosion. It reflects the cost of additional inputs in the future like fertilizer to substitute for topsoil lost in the current year.

The sum of these cost/benefit components represents the net present value to the farmer of choosing the conventional practice for another year. If $\delta_t > 0$, the current profit advantage with conventional tillage outweighs the present value of long-run erosion damage, and the economic incentive is to exploit the soil at least one more year. If $\delta_t < 0$, the cost of long-run erosion damage exceeds the current profit advantage with conventional tillage, and the immediate adoption of conservation is profitable. Expression C-c can be calculated separately in the damage model to estimate the cost of erosion damage as reported for a Palouse site in the main text.

REFERENCE

Walker, D. J. 1982. A damage function to evaluate erosion control economics. Am. J. Ag. Econ. 64(4):690-698.

3

Field Estimates of C Factors: How Good Are They and How Do They Affect Calculations of Erosion?

F. J. Pierce, W. E. Larson, and *R. H. Dowdy*

Techniques for predicting average annual soil erosion have been available since about 1940. The most notable of these has been the Universal Soil Loss Equation (USLE), which has found wide use for the last 25 years. Of the many influences on the degree of erosion, the most important one that individual farmers can control may be the condition of the soil surface and the vegetative cover at the time the potential for erosion by water exists (Renard and Foster, 1983). The coverage and management factor (C factor) of the USLE accounts for this. The ability of the C factor to represent these effects and its influences on the estimate of water erosion is the subject of this paper.

This discussion of C factors considers the definition, context, and calculation of C factors; their importance in the context of conservation policy issues; how reliably and accurately they were used in the 1982 National Resources Inventory (NRI); and their effect on erosion calculations in relation to conservation policy issues.

THE CONTENT AND CALCULATION OF THE C FACTOR

A basic objective of the 1982 NRI was to provide data on the extent and distribution of soil erosion in the United States. As with the 1977 NRI, the USLE was used to estimate soil erosion by water. The USLE was designed to predict long-term average soil losses in runoff from specific field areas in specified cropping and management systems (Wischmeier and Smith, 1978). The USLE has primarily been used to inventory erosion under current conditions and to guide in the development and application of conservation plans (Foster, 1982a).

The USLE is an empirical model that estimates water erosion as a function of six factors, one of which is the C, or cover and management, factor. (For a discussion of the physical factors, RKLS, see Heimlich and Bills, this volume).

The C factor is the ratio of soil loss from an area with specified cover and management to that from an identical area in tilled continuous fallow. It measures the effect of canopy and ground cover on the hydraulics of raindrop impact and runoff; of cover and management on the amount and rate of runoff; of coverage and management on soil structure, organic matter, soil tilth, evapotranspiration, and other soil characteristics; of carryover from previous land use when land use changes; and of roughness from tillage or other disturbances (Foster, 1982a).

These effects are evaluated from soil loss ratios--the ratio of soil loss from a particular practice at a given crop stage on a given soil to that from a unit plot of the same soil. Soil loss ratios vary during the year with crop canopy, ground cover, primary tillage, seedbed preparation, and harvest. A value for C is a weighted soil loss ratio based on the distribution of rainfall erosivity over the year. For cropland, the figures are based on extensive data from natural runoff plots. Ratios from conservation tillage and construction sites are based on data from rainfall simulators. For undisturbed lands such as rangeland and forestland, C factors are based on subfactor relationships for separate effects. The subfactor method for calculating C factors has been described by Wischmeier (1973, 1975).

The C factor is determined by many variables, including weather, that are influenced by management, such as crop canopy, residue mulch, incorporated residues, tillage, and land use residuals. Table 1 illustrates how an annual value for C is calculated (in this case, for continuous corn production). Column 3 lists the cumulative percentage annual erosion index (EI) for the lower peninsula of Michigan; for any given period, this is a numerical measure of the erosive potential of rainfall. Column 5 gives the fraction of EI that occurs during the event in column 1. For example, 39 percent of the annual EI occurs during crop stage 3. The summation of column 7 (the product of columns 5 and 6) is the annual C factor value for continuous corn for this area of Michigan--0.37. Refer to Agriculture Handbook No. 537 (Wischmeier and Smith, 1978) for further details of the calculation.

TABLE 1 Calculation of Annual C Factor for Continuous Production for Central Michigan

Event	Date	Cumulative EI[a] (Percent)	Crop Stage Period[b]	EI in Period (Percent)	Soil Loss Ratio[c]	Crop Stages C Factor
Fall plow	10/15	94	F	18	.44	0.0792
Disk	5/5	12	SB	13	.65	0.0845
Plant corn	5/10	--	--	--	--	--
10% C	6/1	25	1	13	.53	0.0689
50% C	6/20	38	2	15	.38	0.057
75% C	7/10	53	3	39	.20	0.078
Harvest	10/14	92	4L	0	--	--
Fall plow	10/15	92	F	--	--	--
Total						0.37

[a]EI = erosion-times-intensity, a cumulative percentage annual erosion index.
[b]F = rough fallow; SB = seed bed; 1 through 4L = crop stages, as related to canopy development (as percent cover) and harvest.
[c]Incorporates all management variables that affect the C factor.

SOURCE: Calculated using SCS technical guide for Michigan and Wischmeier and Smith (1978).

In general, C value tables are prepared by people experienced in the calculation procedures. Users of the equation then select C factors most appropriate for conditions in the field.

IMPORTANCE OF C FACTORS IN CONSERVATION POLICY

Of the various physical and management factors incorporated in the USLE, the C factor may be the most important, for several reasons. Its range of possible variation affects computed soil loss more than any other USLE factor (Foster, 1982a). C values range from 0.001 for undisturbed forestland with 100 percent cover to 1.0 for clean tilled fallowed land. It is the factor most easily changed through soil management to control erosion. The soil loss ratio for corn in a no-till sod-based system is given in Agriculture Handbook No. 537 as 0.01. Considerable efforts in research and field programs have thus been directed at management practices that affect the C factor.

Public policy concerns have focused in recent years on issues related to the C factor, specifically, conservation tillage practices. The Agricultural Stabilization and Conservation Service (ASCS) currently provides financial assistance to farmers to adopt conservation tillage practices.

Lastly, the C factor is important because it is probably the factor most in need of revision, especially in the areas of the effectiveness of crop residues in controlling erosion (Cogo et al., 1983, 1984; Laflen et al., 1981) and C factors for rangeland (Foster, 1982b; Osborn et al., 1977) and forestlands (Dissmeyer and Foster, 1985).

How Good Are the C Factors?

Over the last 25 years the USLE has evolved as its users gained a better understanding of erosion processes and control. Still, as an estimation procedure the USLE is imperfect and subject to specific limitations. Sources of error include:

- the empirical relationship itself;
- measurement of the parameters that affect the equation parameters;

- the application of the USLE in the field; and
- the application of the equation to situations for which it has not been substantiated.

The latest revision of the USLE was detailed in the Agriculture Handbook No. 537 (Wischmeier and Smith, 1978), which provided the basic guideline for estimating sheet and rill erosion in both the 1977 and 1982 NRIs. Although these guidelines included information on the variables that determine C, they do not reflect current knowledge, for example, on the effectiveness of conservation tillage in erosion control. The context of conservation tillage in relation to C factors is used here to illustrate sources of error in the use of C factors.

Consider errors associated with the empirical relationship itself. The effectiveness of leaving crop residues on the soil surface to control soil erosion is well established. Although the impact of crop residue management is included in the C factor relationship used in the 1982 and 1977 NRIs, more recent information indicates that the importance of residues in erosion control has been underestimated.

C values are currently selected on the basis of tillage system, spring residue weights, and crop residue cover after planting (Laflen et al., 1981). A C value for conventional tillage at a particular crop stage is multiplied by a residue factor based on percentage residue cover. The residue or mulch factor is illustrated in Figure 1. The relationship is described by:

$$F = e^{-bM},$$

where F = the mulch factor; M = residue cover, percent; and b = a coefficient.

The residue cover relationship presented in Agriculture Handbook No. 537 corresponds to a value for b of 0.025. Laflen et al. (1981) reported values for b in the literature ranging from 0.016 to 0.074. Their suggested average value of 0.05 for b is also plotted in Figure 1. Cogo et al. (1984) found b to vary in their study from 0.015 to 0.103. Their findings showed that the b value varied with soil surface roughness and with type and incorporation of residue.

The effect of b on the mulch factor F is measurable (Foster, 1984). At 50 percent residue cover, the mulch factor is 0.5 when b is 0.014 and declines to 0.01 when b equals 0.10. The mulch factor is 0.29 when b is 0.025

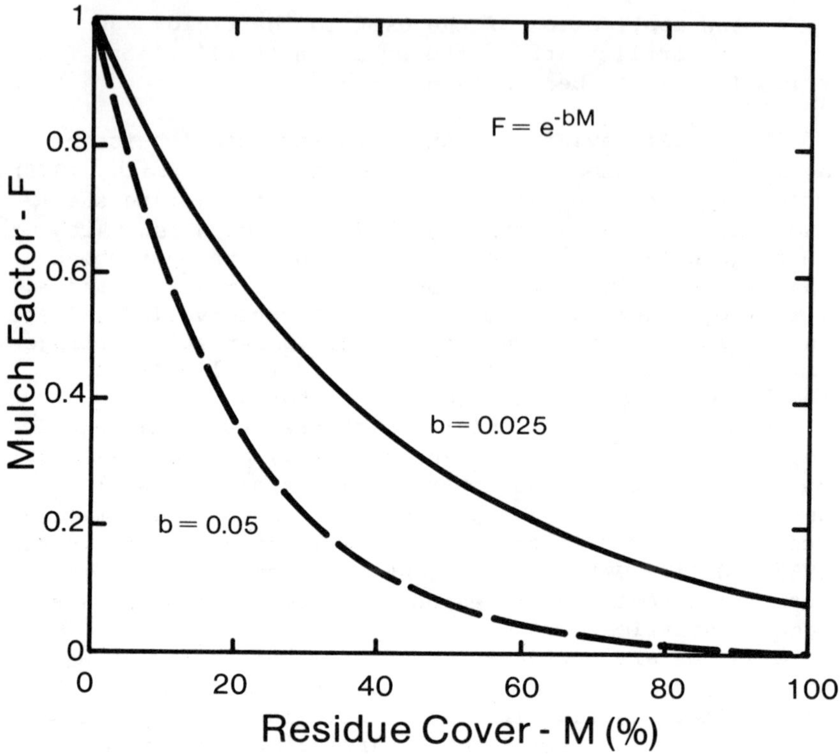

FIGURE 1 The relationship between percentage residue cover (M) and the mulch factor (F) with b of 0.025 and 0.05 (adapted from Laflen et al., 1981).

(Agriculture Handbook No. 537) and is 0.08 when b is 0.05 (Laflen et al., 1981). Two conclusions are clear. First, the effect of residue cover and conservation tillage on erosion control is quite variable. Second, residue cover is more effective in controlling sheet and rill erosion than was considered in the C factors used in the 1977 and 1982 NRIs.

The mulch factor relationship also illustrates a second source of error--that associated with measurement. The importance of the measurement of residue cover becomes clear in the curves in Figure 1. At low levels of residue cover, a small change produces a relatively large change in the mulch factor. Richards et al. (1984) reported on the variation in measurement of residue cover using the line intercept method. Six observers measured residue cover on eight strings placed in a field. The authors reported that the variation among strings was greater than that among observers. The mean residue cover ranged from 26 to 44 percent among strings and from 31 to 43 percent among observers. With a b value of 0.025, the mulch factor would range from 0.52 to 0.33 among strings and 0.46 to 0.34 among observers. Using b of 0.05, the mulch factor ranges from 0.27 to 0.11 and 0.21 to 0.12, respectively.

Colvin et al. (1981) reported the range of residue cover for various tillage systems. For spring tillage, chisel plowing left residue cover that ranged from 40 to 85 percent. Disking in the spring left residues that provided 42 to 73 percent cover. Corresponding mulch factors for b of 0.025 would range from 0.37 to 0.12 for chisel plowing and from 0.35 to 0.16 for disking; for b of 0.05, F ranges from 0.14 to 0.01 and from 0.12 to 0.03, respectively.

Application of the USLE in the field is a third potential source of error. The 1982 NRI was completed over 3 years, and primary sampling points were visited annually throughout the field season. Do the C values in the 1982 NRI accurately reflect the conditions in the field? C values were selected by field personnel from values tabulated for cropping practices in the particular application area based on present practices. Thus, a corn field with a particular production level tilled using a particular conservation tillage system would have a set C factor. A recent study by Peter Nowak (University of Minnesota, personal communication, 1984) of three watersheds in Iowa may reveal a source of error. Of the 200 farmers interviewed, 78 percent claimed to be using conservation tillage in 1982. Yet only 7 percent of the corn acres of those farmers and 26 percent of their soybean acres had the residue cover recommended by the SCS (Soil Conservation Service) to be categorized as conservation tillage. (The recommended rate of corn residue for conservation tillage is 2,000 lbs/acre or more and for soybeans, 1,000 lbs/acre or more.) The implication is that the erosion control attributed to conservation tillage already on the land may be less than indicated in recent estimates of the extent of this tillage.

A fourth source of error is that associated with the application of the USLE to situations for which the equation has not been validated. The USLE has been applied to a wide variety of situations over the years.

Of particular concern has been the quality of C values for undisturbed lands, particularly rangelands (Foster, 1982a,b; Osborn et al., 1977) and forestlands (Dissmeyer and Foster, 1981). The data base used to develop soil loss ratios used to calculate C is mainly for cropland situations (Renard and Foster, 1983).

Rangelands are the largest single land classification in the United States. Osborn et al. (1977) described the problems associated with the application of the USLE to rangeland conditions as expressed in the Walnut Gulch Watershed in Arizona. They identified the determination of the C factor as the greatest uncertainty in the application of the USLE in the Southwest and suggested that when only rangeland vegetation is considered, ground cover is very low and C is very high. Erosion pavement (the concentration of coarse particles at the soil surface resulting from selective erosion of finer particles) present in Walnut Gulch protects the soil from direct raindrop impact and surface runoff erosion and should be considered in erosion estimates. Conversely, although it is valuable as surface protection, the pavement allows runoff to be concentrated between pebbles, thereby increasing erosion Potential. Foster (1982b) questions whether the effects of pavement on erosion potential properly belong in the K factor of the USLE or the C factor. Gullies (channels) apparently play a strong role in sediment yield on even the smallest rangeland watersheds, a factor not considered in the USLE. Osborn et al. (1977) suggest that a possible channel factor (Ec) be included in the USLE.

The fact that single storm events can dominate soil loss from rangelands presents a major consideration when applying the USLE to them. Trieste and Gifford (1980) assessed the applicability of the USLE to rangelands on a per-storm basis and concluded that where sediment yields are dominated by single storm events, the USLE does not explain soil loss and may give misleading rather than useful results. While Foster et al. (1981) disagree with Trieste and Gifford's conclusions, the applicability of the USLE to rangelands remains an issue.

Dissmeyer and Foster (1981) used the subfactor approach to develop a procedure for estimating C factors for forest conditions in the southeastern United States and, more recently (Dissmeyer and Foster, 1985), discussed the application of the USLE to other forest, range, and wildland conditions where data to develop C factors are limited or unavailable. This new information was not available in the 1982 or 1977 NRIs, but will certainly be included in future estimates. As the USLE is extended to new situations and reflects new information, however, various NRIs may not be comparable.

C FACTORS IN THE 1982 NRI

Data compiled in the 1982 NRI provide some information as to the reliability of the C factors. Data from the 1982 NRI were summarized nationally and for four Major Land Resource Areas (MLRAs) in the United States (USDA, 1981). Table 2 lists this summary data on land use, potential for erosion, and C factors nationally and for the four MLRAs.

The distribution of C factors for cropland identified in MLRA 105 as cropped to corn for the 4 years reported in 1982 is given in Figure 2. As would be expected, the majority of sampling points without conservation tillage practices falls in a single class, 0.35 to 0.40, and most of those with conservation tillage are in the class 0.15 to 0.20. The broad range of values under each category and the occurrence of a substantial number of sampling points with C factors in the lower classes was unexpected, given that during the last 4 years the crop was corn. The higher C values for conservation tillage might be explained in terms of differences in tillage practice and residue management. Low C values for cropland not in conservation tillage are surprising and warrant further investigation.

Variation in C values with potential for erosion may be indicative of the effectiveness of conservation practices. On a national basis, the average C factor varied little with land's inherent potential for erosion, as estimated by the RKLS product of the USLE (see Figure 3). These data suggest that, on the average, conservation tillage practices have not been used on soils with high erosion potentials as much as they could be. The trends vary with region (see Figure 4). For MLRA 105, C values were quite low compared with other regions and decreased significantly with increased erosion potential. This, combined with a similar decrease in P factor with erosion potential, resulted in a diminished slope (0.13) of the estimated versus potential erosion line, relative to other areas. Little change occurred in the C values in MLRAs 103, 134, and 136, and the slope of this line was considerably higher for these areas (0.29, 0.25, and

TABLE 2 U.S. Cropland and Four Major Land Resource Areas: Distribution of Acres and Percentage with RKLS <10 and <20 Tons/Acre/Year, and C Factors

MLRA	Cropland (Million Acres) Row/Close-Grown Crops	Other	Potential	Acres in Row/Close-Grown Crops with RKLS: <10	<20	Weighted Average C Factor
103	13.8	0.6	0.6	70%	87%	0.383
105	3.8	2.0	1.1	11	23	0.205
134	7.1	0.3	2.9	4	48	0.294
136	3.4	0.8	7.3	4	19	0.311
National	323.2	86.3	154.4	52	76	0.300

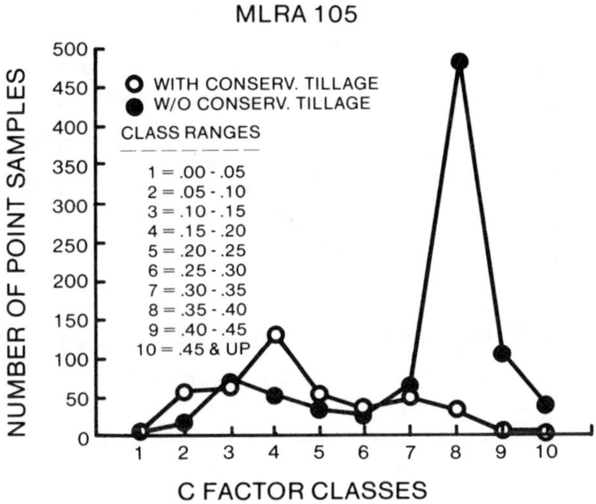

FIGURE 2 The distribution of C factors for land cropped to 4 years of corn in MLRA 105 expressed in terms of the number of point samples within a range of C values as summarized from the 1982 NRI.

0.29, respectively). Nationally, the slope of this line was 0.22.

Table 3 summarizes acreage and C factors nationally by crop. The major crops (corn, soybeans, wheat, and cotton) account for 82 percent of the 323 million acres in row and close-grown crops. The average C factors were high relative to that obtainable with conservation tillage.

The potentials for erosion, as indicated by the RKLS product of the USLE, vary with crop (see Table 3) and show that nationally a considerable portion of cropland soils have a low potential for erosion. This seems especially true for cotton and may explain the high C values for land planted in this crop. Soils with low potential for water erosion, however, may be highly susceptible to wind erosion. Since the C factor, high C factors may indicate a low degree of protection from wind.

Table 4 gives the potential for erosion and the percentage of land in conservation tillage by crop for the four MLRAs. About 70 percent of the cropland in MLRA 103 had a potential for sheet and rill erosion of less than 10 tons/acre/year. Of all cropland in conservation tillage in MLRA 103, some 65 percent was on land with

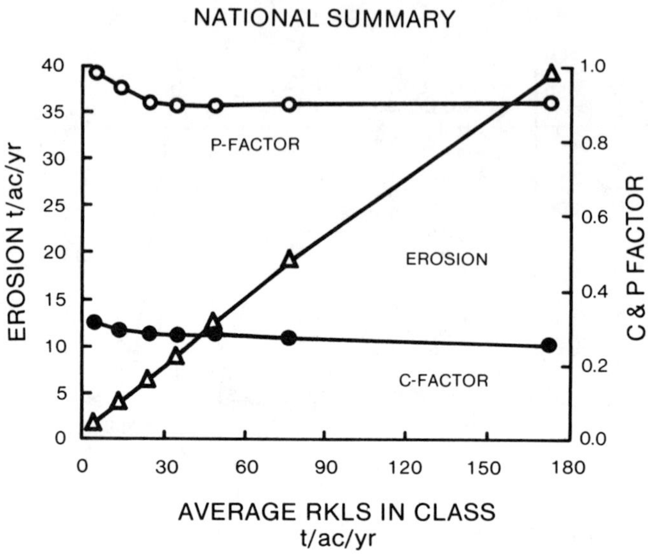

FIGURE 3 Plots of the 1982 NRI weighted average erosion rate (tons/acre/year), C factor, and P factor versus the potential for erosion (tons/acre/year) as expressed by the RKLS product of the USLE. Data are summarized nationally.

RKLS of less than 10, and an additional 20 percent was on cropland with an RKLS of 10 to less than 20 tons/acre/year. For MLRAs 105, 134, and 136, however, 77, 52, and 81 percent of the cropland had an erosion potential of over 20 tons/acre/year, with 24, 17, and 12 percent of that land, respectively, being in conservation tillage. This does not suggest that conservation tillage is not effective on land with low water erosion potential. Table 5 shows that for land with a low potential for erosion (RKLS less than 10 tons/acre/year) in MLRA 103, wind erosion increases as the C factor rises, indicating that conservation measures are effective in controlling wind erosion. However, wind erosion rates are much lower than water erosion rates on sloping lands. There is no apparent explanation for higher wind erosion rates under conservation tillage. The wisdom of considering wind and water erosion estimates cumulatively, which is so often done, is called into question here.

The Land Capability Class System (Klingebiel and Montgomery, 1961) is extensively used by the SCS and is

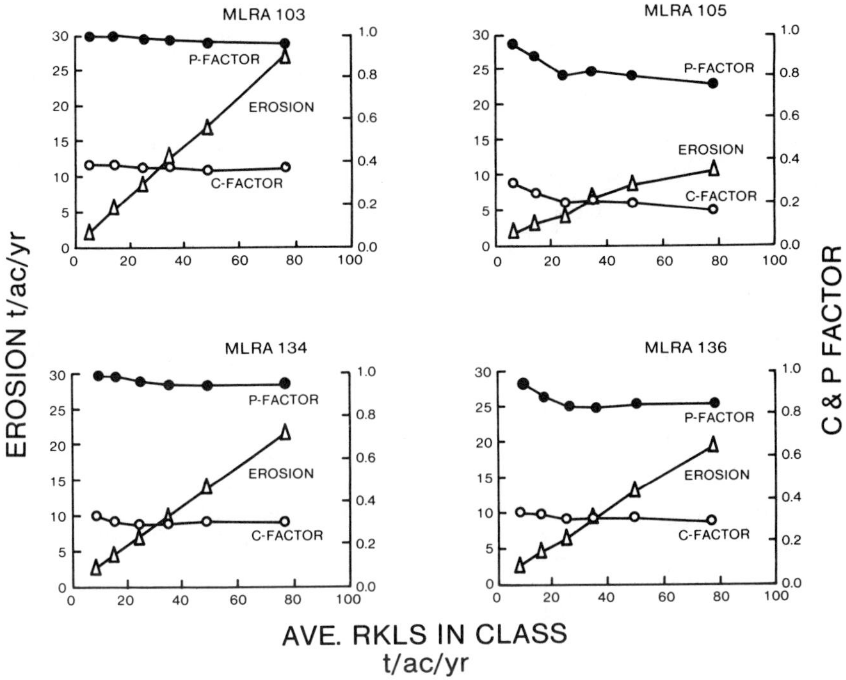

FIGURE 4 Plots of the 1982 NRI weighted average erosion rate (tons/acre/year), C factor, and P factor versus the potential for erosion (tons/acre/year) as expressed by the RKLS product of the USLE for MLRAs 103, 105, 134, and 136.

included in the 1982 NRI. One subclass of this system, e or erosion subclass, identifies lands for which erosion is a severe limitation to land use. Directing conservation practices at this land should be reflected in its C values.

The change in C factor with increase in RKLS for selected land capability classes is given in Figure 5. Nationally, class I land showed the highest C factors when potential erosion was less than 60 tons/acre/year and the lowest C factors when RKLS exceeded that level. There was little change in C for class IIe land. Class IIIe and IVe land showed declines in C at lower potential erosion and a slight increase in C with further increases in RKLS. It appears that conservation measures are considered important on the more erosive class I land and less important on the intermediate erosive class I land

TABLE 3 U.S. Population and Actual Cropland: Distribution of Acres and Percentage with RKLS <10 and <20 Tons/Acre/Year, and C Factors

Crop	Cropland (Million Acres)	Percentage of Acres with RKLS		Weighted Average C Factor
		<10	<20	
Corn	91	48	69	0.31
Soybeans	67	41	71	0.34
Cotton	17	71	91	0.48
Sorghum	17	52	76	0.29
Wheat	89	57	80	0.24
Potential cropland	154	41	62	NA
Total cropland	409	53	75	0.27

SOURCE: 1982 NRI.

TABLE 4 Four Major Land Resource Areas: Distribution of Acreage by Crop, Potential Erosion Class, and Conservation Tillage

Crop	Cropland (1,000 Acres)	Conservation Tillage (Percentage of Acres)	Potential Erosion Class (Percentage of Acres with RKLS of Tons/Acre/Year)[a]		
			<10	10-20	>20
MLRA 103					
Corn	7,162	14	69 (13)	18 (16)	13 (18)
Soybeans	5,854	13	72 (12)	16 (16)	12 (16)
Wheat	388	10	76 (11)	17 (7)	8 (3)
Cropland	14,443	13	70 (12)	17 (15)	13 (16)
MLRA 105					
Corn	3,166	28	13 (19)	15 (25)	72 (30)
Soybeans	198	15	31 (5)	21 (19)	47 (20)
Wheat[b]	--	--	--	--	--
Cropland	5,820	23	11 (14)	12 (23)	77 (24)
MLRA 134					
Corn	445	21	6 (14)	29 (15)	65 (24)
Soybeans	4,293	16	4 (9)	44 (13)	52 (19)
Wheat	552	23	2 (12)	33 (21)	65 (25)
Cotton	860	4	7 (3)	43 (7)	50 (3)
Cropland	7,469	14	4 (7)	44 (13)	52 (17)
MLRA 136					
Corn	804	12	7 (8)	20 (20)	73 (10)
Soybeans	1,149	18	3 (12)	15 (11)	82 (20)
Wheat	744	18	3 (56)	14 (11)	83 (18)
Cropland	4,223	12	4 (17)	15 (11)	81 (12)

[a]Numbers in parentheses are percentage of acres in conservation tillage.
[b]Number of sampling points too small.

SOURCE: 1982 NRI.

TABLE 5 Major Land Resource Area 103: Acreage with Low Potential Erosion, and Wind Erosion Rates by Conservation Tillage Use

C Class	Potential Erosion (1,000 Acres with RKLS <10 Tons/Acre/Year)	Wind Erosion Rates (Tons/Acre/Year)		
		Conservation Tillage Used	No Conservation Tillage	Overall
<0.1	162	0.9	0.8	0.7
0.1--<0.2	463	3.7	1.8	1.6
0.2--<0.3	911	4.2	3.3	3.3
0.3--<0.4	3,741	4.4	3.9	3.7
0.4+	4,798	5.6	6.1	6.0

SOURCE: 1982 NRI.

(RKLS 30 to 60 tons/acre/year). For land in the erosive subclasses IIe, IIIe, and IVe, conservation practices are not emphasized on lands with greater erosion potential. Although Figure 5 shows that a large portion of subclass e land had a low potential for water erosion, a significant portion of the land with high erosion potential is not receiving conservation treatments.

As earlier indicated, the situation varies by area. (see Figure 6). For MLRA 103, C was high (0.51) for class I land with high erosion potential (although the acreage was small). There were over a million acres of class IIIe land in this area in row and close-grown crops (7 percent of the total), most of which had a high potential for erosion. The C factor averaged 0.37 for class IIIe land in MLRA 103 and did not decrease with increase in RKLS.

C factors for MLRA 105 were low for row and close-grown crops, averaging 0.21, and were lowest for IIIe and IVe land. They tended to decrease even further with increasing erosion potential. Class IIe, IIIe, and IVe land accounted for 70 percent of the land in row and close-grown crops in MLRA 105.

MLRA 134 showed a linear decrease in C with increasing potential for erosion for class I land in row and close-grown crops. C factors for class IVe land started high but decreased with increasing erosion potential for RKLS less than 40 tons/acre/year. C factors for high-erosion-potential class IVe land and for class IIe and IIIe land did not decline with increasing RKLS and averaged about 0.30.

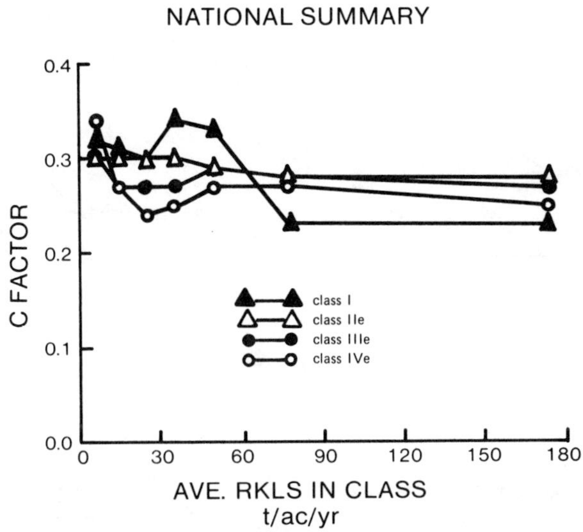

FIGURE 5 Plots of the weighted average C factor versus potential for erosion (tons/acre/year) as expressed by the RKLS factor of the USLE for land capability subclasses I, IIe, IIIe, and IVe. Data were summarized nationally from the 1982 NRI.

In MLRA 136, class IVe cropland with medium potential for erosion (RKLS 20 to 40 tons/acre/year) had lower C factors than cropland in other classes. For the most part, C factors for land in row and close-grown crops in this area did not change with increasing erosion potential.

There is no one pattern in the use of conservation measures on class I cropland. Farmers consistently did not increase their use of conservation measures in proportion to increasing erosion potential, especially on lands in the eroded subclasses. This can be a substantial problem considering the acreage in capability classes IIe, IIIe, and IVe (see Figure 7). Both nationally and in MLRA 103, class IIIe land is the primary problem, while in MLRAs 105, 134, and 136, class IIe as well as class IIIe land in row and close-grown crops is important.

The C values in the 1982 NRI indicate three things: (1) a considerable portion of the land in conservation tillage has a low potential for water erosion, (2) conservation tillage practices are not adequate on land with medium to high potential for water erosion, and (3) some C values in the 1982 NRI are unexpected and suspect.

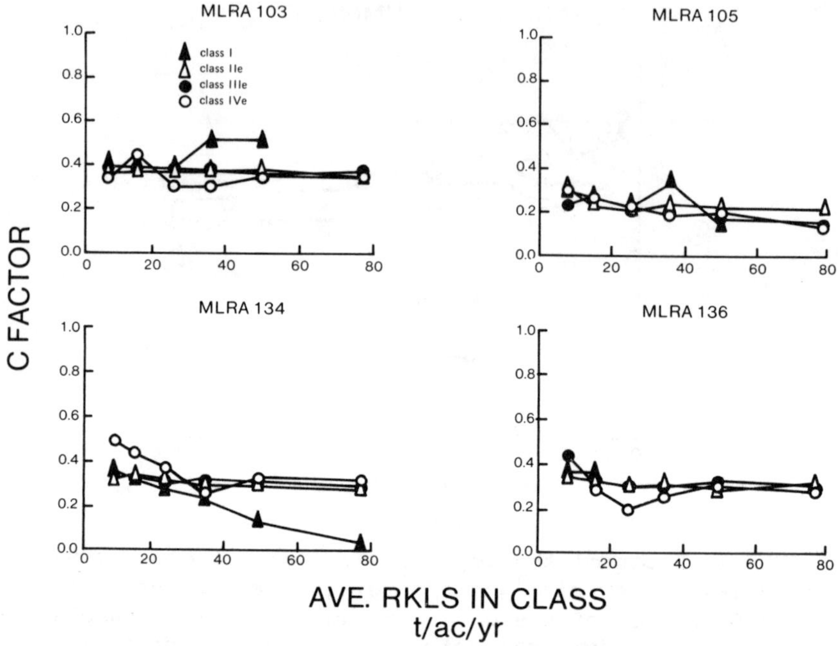

FIGURE 6 Plots of the weighted average C factor versus potential for erosion (tons/acre/year) as expressed by the RKLS factor of the USLE for land capability subclasses I, IIe, IIIe, and IVe for MLRAs 103, 105, 134, and 136.

The Effect of C Factors on Calculation of Erosion

Given the the current state of knowledge about the USLE, what effect would reducing C on cropland have on the extent and degree of soil erosion by water? In a related question, what would be the effect on water erosion of bringing potential cropland into production under various management practices?

Figure 8 plots the percentage of acres nationally with erosion rates as estimated with the USLE under three assumed C factors for land currently in row and close-grown crops, land with a high potential for cropland conversion, and land with a medium potential for conversion. With a C factor of 0.3, 73 percent of the land with row and close-grown crops would have erosion rates under 5 tons/acre/year. This compares with the 75 percent actually estimated in 1982 with a calculated

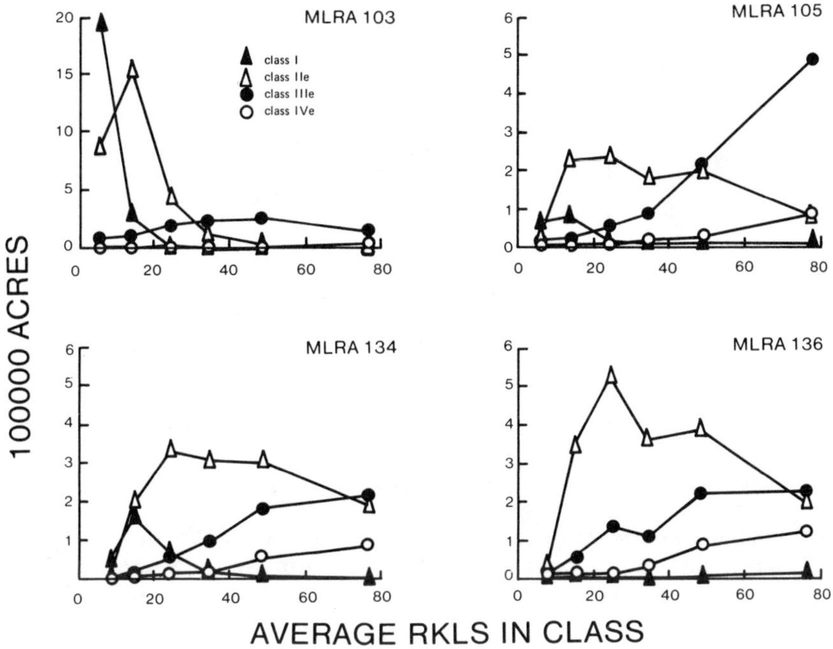

FIGURE 7 Distribution of land in land capability subclasses I, IIe, IIIe, and IVe by RKLS class for MLRAs 103, 105, 134, and 136.

average C factor of 0.3 for row and close-grown crops (see Table 1). With an assumed C factor of 0.1, 93 percent of U.S. acreage in row and close-grown crops would have eroded less than 5 tons/acre/year. For land with a high potential for conversion to cropland, the comparable figures are 60 and 89 percent, respectively. And for land with a medium conversion potential, the values are 57 and 82 percent, respectively. These numbers suggest that some latitude exists for reducing erosion through conservation on both existing and potential cropland. Some 7 percent of the cropland would require additional erosion control practices.

Again, the situation varies considerably throughout the country (see Figure 9). The situation in MLRA 103 is similar to that nationally. But in MLRA 105, the erosion situation is improved by increased use of conservation practices. However, the erosion potential is so great for these soils that practices such as conservation tillage are not enough to solve the erosion problem. The

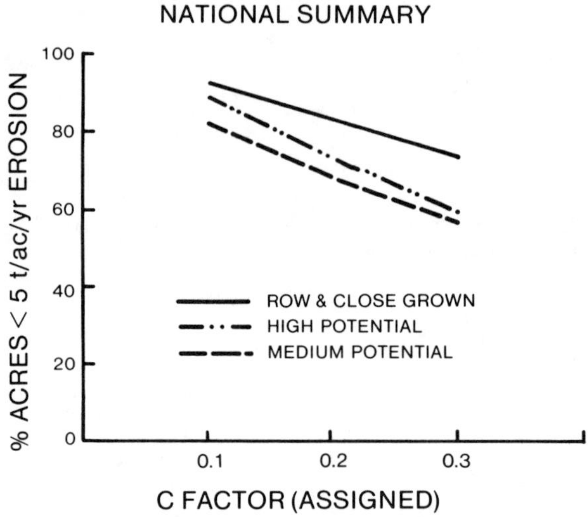

FIGURE 8 Percentage of acres nationally with USLE erosion rates <5 tons/acre/year at assumed levels of C factors for land in row and close-grown crops in 1982.

data also suggest that potential cropland from this area is less desirable than that from MLRA 103 when viewed from strictly an erosion perspective. Land in MLRA 134 is intermediate between MLRAs 103 and 105. Notice, however, the steepness of the curves. This suggests that the benefits of conservation tillage systems on these soils should be dramatic and very visible. The situation for MLRA 136 is similar to MLRA 105 in position and to MLRA 134 in terms of slopes. In all cases, land that has a medium potential for cropland conversion is always below current cropland and high potential land, and the curve generally slopes less.

It is clear that reducing the C factor through management practices can significantly affect soil erosion. But erosion control measures beyond conservation tillage need to be explored and promoted on the land.

CONCLUSIONS

Foster et al. (in press) described the USLE as "the world standard for an equation to estimate sheet and rill erosion," saying that "no other current equation or procedure for estimating erosion approaches, as a whole,

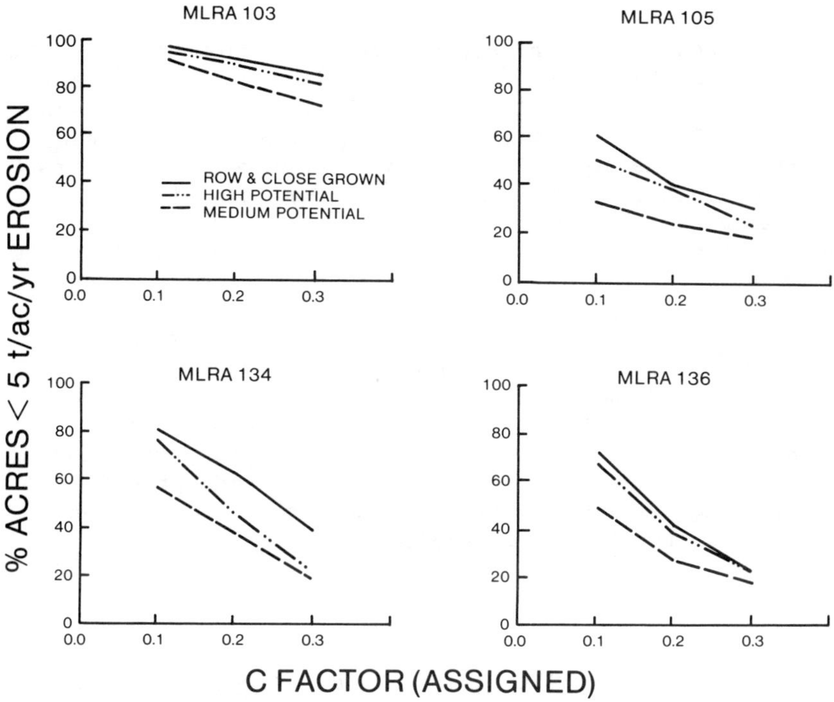

FIGURE 9 Percentage of acres in MLRAs 103, 105, 134, and 136 with USLE erosion rates >5 tons/acre/year at assumed levels of C factors for land in row and close-grown crops in 1982.

the USLE in ease of application, breadth of application, and accuracy."

What can be concluded from this brief look at the reliability and accuracy of the C factor in the USLE? A few things are apparent. First, conservation tillage may be more effective in controlling erosion than previously considered. This would support public policies that promote the use of conservation tillage to control erosion. Second, conservation tillage is currently concentrated on land with low potential for erosion. Third, there is probably less crop residue management on the soil surface than recent data on the extent of conservation tillage imply. From a policy standpoint, these two items would suggest that technology transfer or the extension of information regarding conservation tillage to the land user is not adequate and that more effort needs to be directed toward farmers who have land with medium and high potential for erosion.

Fourth, conservation tillage is not the sole solution for all soils and landscapes needing erosion control. It is critical that efforts not be limited to a few select control measures. Fifth, C factors will be improved in future NRIs. Revisions of the USLE are currently being done at the National Soil Erosion Laboratory. Last, as C factors change, comparisons to earlier NRIs will become complicated. There are often attempts to compare NRIs in hopes of discerning a change in soil erosion attributable to conservation policy. A measure of caution should accompany such endeavors.

In closing, it is far easier to criticize than to construct. The USLE has been an important tool and is reliable when data are available and the equation has been evaluated. Erosion technology has been greatly advanced under the umbrella of the USLE (Dissmeyer and Foster, 1985). Erosion prediction will undoubtably change and improve as the knowledge to do so is gained, in part to the credit of the USLE.

REFERENCES

Cogo, N. P., W. C. Moldenhauer, and G. R. Foster. 1983. Effect of residue, tillage-induced roughness, and runoff velocity on size distribution of eroded soil aggregates. Soil Sci. Soc. Am. J. 47:1005-1008.

Cogo, N. P., W. C. Moldenhauer, and G. R. Foster. 1984. Soil loss reductions from conservation tillage practices. Soil Sci. Soc. Am. J. 48:368-373.

Colvin, T. S., J. M. Laflen, and D. G. Erbach. 1981. A review of residue reduction by individual tillage implements. Pp. 102-110 in Crop Production with Conservation in the '80s. Proc. Am. Soc. Agric. Eng., December 1-2, 1980. Publ. No. 7-81. St. Joseph, Mich.: American Society of Agricultural Engineers.

Dissmeyer, G. E., and G. R. Foster. 1981. Estimating the cover-management factor (C) in the Universal Soil Loss Equation for forest conditions. J. Soil Water Conserv. 36:235-240.

Dissmeyer, G. E., and G. R. Foster. 1985. Modifying the Universal Soil Loss Equation for new situations. In Proc. of the Second International Conference on Soil Erosion and Conservation, Honolulu, Hawaii, January 16-23, 1983. Ames, Iowa: Soil Conservation Society of America.

Foster, G. R. 1982a. Special problems in the application of the USLE to rangelands: C and P factors. Pp. 96-100 in Proc. of Workshop on Estimating Soil Erosion and Sediment Yield from Rangelands. ARM-W-26. Washington, D.C.: USDA Agricultural Research Service.

Foster, G. R., 1984. Plan for revising Agriculture Handbook 537 to update the Universal Soil Loss Equation (USLE). Unpublished, April 15, 1984.

Foster, G. R., J. R. Simanton, K. G. Renard, L. J. Lane, and H. B. Osborn. 1981. Discussion of "Application of the Universal Soil Loss Equation to rangelands on a per-storm basis," by Trieste and Gifford in Journal of Range Management 33:66-70, 1980. J. Range Manage. 34:161-165.

Foster, G. R., J. M. Laflen, and C. V. Alonso. In press. A replacement for the Universal Soil Loss Equation (USLE). In Proc. of the ARS-SCS Natural Resources Modeling Workshop. Washington, D.C.: USDA Agricultural Research Service.

Klingebiel, A. A., and P. H. Montgomery. 1961. Land Capability Classification. Agriculture Handbook No. 210, USDA Soil Conservation Service. Washington, D.C.: U.S. Government Printing Office.

Laflen, J. M., W. C. Moldenhauer, and T. S. Colvin. 1981. Conservation tillage and soil erosion on continuously row-cropped land. Pp. 121-133 in Crop Production with Conservation in the '80s. Proc. Am. Soc. Agric. Eng., December 1-2, 1980. Publ. No. 7-81. St. Joseph, Mich.: American Society of Agricultural Engineers.

Osborn, H. G., J. R. Simanton, and K. G. Renard. 1977. Use of the Universal Soil Loss Equation in the semiarid Southwest. Pp. 41-49 in Soil Erosion: Prediction and Control. Special Publ. No. 21. Ankeny, Iowa: Soil Conservation Society of America.

Renard, K. G., and G. R. Foster. 1983. Soil conservation: Principles of erosion by water. Pp. 155-176 in Dryland Agriculture. Agronomy Monograph No. 23. Madison, Wis.: American of Society of Agronomy.

Richards, B. K., M. F. Walter, and R. E. Muck. 1984. Variation in line transect measurements of crop residue cover. J. Soil Water Conserv. 39:60-61.

Trieste, D. J., and G. F. Gifford. 1980. Application of the Universal Soil Loss Equation to rangelands on a per-storm basis. J. Range Manage. 33:66-67.

USDA Soil Conservation Service. 1981. Land resource regions and major land resource areas of the United States. Agriculture Handbook No. 296. Washington, D.C.: U.S. Government Printing Office.

Wischmeier, W. H. 1973. Conservation tillage to control water erosion. Pp. 133-144 in Proc. of the National Conservation Tillage Conference, Des Moines, Iowa, March 28-30, 1973. Ankeny, Iowa: Soil Conservation Society of America.

Wischmeier, W. H. 1975. Estimating the soil loss equation's cover and management factor for undisturbed areas. Pp. 118-124 in Present and Prospective Technology for Predicting Sediment Yields and Sources. USDA Agricultural Research Service. ARS-S-40. Washington, D.C.: U.S. Government Printing Office.

Wischmeier, W. H., and D. D. Smith. 1978. Predicting Rainfall Erosion Losses: A Guide to Conservation Planning. Agriculture Handbook No. 537, USDA Science and Education Administration. Washington, D.C.: U.S. Government Printing Office.

Discussion

William C. Moldenhauer

The National Soil Erosion Laboratory is very much aware of the problems that Pierce and Larson pose concerning the accuracy of erosion measurement and effectiveness of cropping and management, especially with the added complication of surface residue. Many of the problems cited can be corrected by building a sufficient data base. Unfortunately, the field is not moving as fast as it should to build this data base. Considerably more research is needed on the roughness-cover-soil type interaction. The soil erosion laboratory's work and the tillage research conducted by John Laflen of the Agricultural Research Service at Ames, Iowa, certainly imply that residue is more effective in controlling erosion than USDA Agriculture Handbook No. 537 shows. But much more data are needed before it can be said with confidence that these numbers can be extrapolated over a wide area. Also, concentrated flow cuts the erosion control effectiveness of mulch drastically by either undercutting or floating the mulch away. The National Soil Erosion Laboratory has done some work on this, but more is needed.

The percentage of residue left at planting time on tilled land in the Corn Belt is certainly very disappointing. The percentage of mulch cover left for overwintering is quite good on many fields, but by planting time it has been reduced to the point of ineffectiveness in many, if not most, cases.

Field measurement of mulch cover is not an exact science at the present time, as Pierce and Larson point out. There is a great deal of human error involved. The National Soil Erosion Laboratory is working with an image analyzer to try to perfect a standard against which measuring techniques can be compared. Lowery et al. (1984) compare different techniques. A standard for practice using the technique is essential, and goes a long way toward making estimates more uniform. However, an accurate standard is essential, or there is a tendency to become biased toward the leader, whose estimates may or may not be the most accurate of the group.

Adding a channel factor to the USLE (Universal Soil Loss Equation) is being investigated, because of the concentrated flow erosion or ephemeral gullying common on cropland (see Barfield; Foster, this volume). This investigation may help in applying a channel factor on rangeland. A decision must be made on how to predict erosion with consideration of surface stoniness, a condition that has been encountered in Indiana on reclaimed strip-mined soils. There are many such areas in the United States and elsewhere. Box (1981) has discussed the effect on erosion of surface stoniness on cropland soils in the U.S. Southeast. Collinet and Valentin (1984), in West Africa, found C to go from 0.52 with 5 percent fragment cover to 0.005 with 80 percent cover.

It is difficult to visualize C factors of 0.05 to 0.30 with moldboard-plowed continuous corn unless the plowing is extremely rough. It is much easier to visualize high C values with conservation tillage because of the very broad definition of this type of tillage. These values should be examined for an explanation.

It also seems unlikely that wind erosion is higher with conservation tillage unless residue cover is minimal and flat. In this case, surface roughness caused by the moldboard plow may be more effective than this sparse, flat residue. Fryrear has described this situation in the High Plains of Texas (Moldenhauer et al., 1983).

It was surprising to find virtually no decrease in the C factor [except in MLRA (Major Land Resource Area) 105] as RKLS increased. This seems to reflect a much lower C factor than is necessary for erosion control on the low RKLS situations and that conservation tillage is being done for reasons other than erosion reduction. This is a very significant finding of this study and shows that professional conservationists may be taking more credit than they deserve for reducing erosion through conservation tillage. Much work must be done to convince people to use conservation tillage as an erosion reduction measure on erodible land. But, just as a great breakthrough occurred in the use of conservation tillage when all farmers in the Corn Belt bought a chisel plow to work their soybean land, many then found they could also use it on corn stalks. If farmers with low RKLS (erosivity/erodibility) situations can make conservation tillage work and are using it, it will be much easier to convince farmers with high RKLS (erosivity/erodibility) situations to use these systems too.

Farm operators adopt conservation practices for a variety of reasons. Pierce and Larson should be applauded for looking at the big picture and showing the difficulties of applying effective conservation to the land. This helps to focus research and education efforts where they will be most effective.

The impatience young scientists feel with the slow pace of solving some of our erosion control and estimation problems is understandable. Until as recently as 10 to 15 years ago, very few farmers would even talk about changing their cropping systems and tillage practices. Yet much progress has been made in 30 years, due to the USLE and early tillage research. Real appreciation of this equation comes only with the perspective engendered by a time when it was impossible to put any quantitative value on soil erosion.

Research has been brought to current levels by pioneers in the area of erosion prediction--Dwight Smith, George Browning, Austin Zingg, Walter Wischmeier, and G. W. Musgrave--and by W. E. Larson's pioneering paper on tillage modeling in 1964. Much of the basis of these early efforts was expert intuition. Without these early efforts, scientists might still be stumbling around with no suitable quantitative estimates of any of the factors involved in soil erosion or tillage. Today's modelers are building on these efforts with tools undreamt of in the 1940s and 1950s. Progress is being made faster than anyone could have thought possible just a few years ago. But expert intuition should not be ignored: It is what scientists used in place of computer models, and it is still essential today.

REFERENCES

Box, J. E., Jr. 1981. The effects of surface slaty fragments on soil erosion by water. Soil Sci. Soc. Am. J. 45:111-116.

Collinet, J., and C. Valentin. 1984. Evaluation of factors influencing water erosion in West Africa using rainfall simulation. Challenges in African Hydrology and Water Resources. Proc. of the Harare Symposium, Publ. No. 144. Difford, Wallingford, England: International Association of Hydrological Sciences.

Larson, W. E. 1964. Soil parameters for evaluating tillage needs and operations. Soil Sci. Soc. Am., Proc. 28:118-122.

Lowery, B., T. M. Lillesand, D. H. Mueller, P. Weiler, F. L. Scarpace, and T. C. Daniel. 1984. Determination of crop residue cover using scanning microdensitometry. J. Soil Water Conserv. 39:402-403.

Moldenhauer, W. C., G. W. Langdale, W. Frye, D. K. McCool, R. I. Papendick, D. E. Smika, and D. W. Fryrear. 1983. Conservation tillage for erosion control. J. Soil Water Conserv. 38:114-151.

4

Understanding Ephemeral Gully Erosion

G. R. Foster

CLASSICAL FORMS OF EROSION

The classical forms of erosion by water that occur within farm fields are sheet, rill, and gully erosion (Hutchinson and Pritchard, 1976). Sheet erosion, a uniform removal of soil, is almost imperceptible, although rates as high as 20 tons/(acre • year) have been measured (Meyer, 1981). Erosion of this magnitude is usually considered to be more than the soil can tolerate without serious degradation (Schertz, 1983). Rill erosion is defined as erosion in numerous small channels that can be obliterated by tillage (Hutchinson and Pritchard, 1976). Although sheet erosion is not obvious, rills, typically about 6 inches wide and 4 inches deep, are very obvious. They can follow tillage marks, or they may develop much like a drainage network of rivers in a large basin. Severe rill erosion can exceed 200 tons/acre • year).

According to the modern theory of rill-interrill erosion (Foster and Meyer, 1975), flow concentrates in many small downslope channels that are uniformly distributed across most landscapes, and it is part of overland flow. Any erosison that occurs on these areas is called rill erosion. Spaces between the rills are called interrill areas, and erosion on them is called interrill erosion. Raindrops detach soil particles on interrill areas, and thin flow, enhanced by the raindrop impact (Moss et al., 1979), moves the sediment laterally to the rill areas (Foster, 1982b), where most downslope transport of sediment occurs (Foster and Meyer, 1972b).

Interrill and rill erosion, which are combined in erosion estimates from the Universal Soil Loss Equation (USLE) (Wischmeier and Smith, 1978), are usually considered together when assessing the impact of erosion on farm fields. Since tillage obliterates rills each year, interrill and rill erosion remove soil uniformly in a local sense, although erosion varies greatly over the landscape. For example, the maximum erosion rate on a typical complex land profile can be five times the average rate for a uniform profile on the average steepness of the complex profile (Perrens et al., 1985). Values for interrill and rill erosion, listed as sheet and rill erosion in the National Resources Inventories (NRI) done by the U.S. Department of Agriculture (USDA), are generally considered to be estimates of erosion at a sample point on the landscape, which is not strictly true (<u>see</u> Appendix A).

Classical gully erosion is defined as erosion in channels that are too deep to cross with farm equipment (Hutchinson and Pritchard, 1976). Once established, gullies are permanent unless they are filled with soil moved with heavy equipment. Gullies remove portions of fields completely from production, and they divide fields, which reduces the efficiency of large farm equipment. Obviously, gully erosion significantly reduces land quality and value.

Gullies often develop from intense erosion caused by flow over a steep overfall at the head of the gully. This overfall, called a headcut, moves upstream in a natural drainageway, and it can be initiated offsite and move into a field. Gullies can also be enlarged by lateral erosion, sloughing of their sidewalls, and clean-out of debris by flow in the gullies. Subsurface flow through the gully walls can significantly reduce soil strength and accelerate gully erosion (Piest et al., 1975a).

A NEW TYPE--EPHEMERAL GULLY EROSION

Soil conservationists have recently noted that an important erosional area and source of sediment within fields is being overlooked (Foster, 1982a). Among other terms, it has been called ephemeral gully erosion, concentrated flow erosion, and megarill erosion. The topography of most fields causes runoff to collect and concentrate in a few major natural waterways or swales before leaving the fields (Foster, 1982a; Thorne, 1984). The erosion that occurs in these channels is what is known as ephemeral gully erosion.

These channels are the main drainage system for a field, and most water and sediment are discharged from fields through them. A single branch of this channel network has a major effect on water and sediment delivery from a field, whereas a single rill is one of many and has little effect by itself on the total hydrologic and erosional response of a field. Flow in rills is usually classified as a part of overland flow that is assumed to occur uniformly across a slope even though it is concentrated in rills. In contrast, flow in ephemeral gully areas is clearly channelized (Foster, 1982a). Ephemeral gullies recur in the same area each year; new rills, on the other hand, are strongly influenced by tillage marks and often are reformed in new locations from year to year (Foster et al., in press). Ephemeral gully areas within fields are plowed in and tilled across annually, in contrast to the permanency of classical gullies. Therefore, an ephemeral gully is short-lived, since the area is restored annually by tillage (hence the name ephemeral gully erosion).

Table 1 lists the characteristics of the three types of erosion caused by flow within fields--rill erosion, ephemeral gully erosion, and classical gully erosion. In principle, erosion in each of these eroded channels is by concentrated flow, and therefore several of the erosional processes are the same for each type.

USLE estimates include rill erosion, but the equation clearly does not encompass ephemeral gully erosion. The USLE was empirically derived from plot data where typical slope lengths were 36, 73, and 145 feet, except for two studies where the maximum slope lengths were 270 and 630 feet (Foster, 1982c). In all cases, the largest eroded channels were rills. Also, the USLE includes interrill erosion caused by raindrop impact, while ephemeral erosion is caused entirely by flow. "Defined flow channels," according to the USLE slope length definition, mark the end of USLE slope lengths (Wischmeier and Smith, 1978); "defined channels" include ephemeral gully areas even if no erosion occurs in them.

No prediction method similar to the USLE is available for estimating ephemeral gully erosion, although such an equation is needed. Some governmental assistance programs, such as the Agricultural Conservation Program (ACP), require that erosion reduction from conservation practices (including terraces, waterways, and other similar structural practices that control ephemeral gully erosion) be estimated. Nonpoint source pollution and

TABLE 1 Characteristics of Rill Erosion, Ephemeral Gully Erosion, and Classical Gully Erosion

Rill Erosion	Ephemeral Gully Erosion	Classical Gully Erosion
Rills are normally erased by tillage; they usually do not recur in the same place	Ephemeral cropland gullies are temporary features, usually obscured by tillage; recur in the same location	Gullies are not obscured by normal tillage operations
May be of any size but are usually smaller than ephemeral cropland gullies	May be of any size but are usually larger than rills and smaller than permanent gullies	Usually larger than ephemeral cropland gullies
Cross sections tend to be narrow relative to depth	Cross sections tend to be wide relative to depth; sidewalls frequently are not well defined; headcuts are usually not readily visible and are not prominent because of tillage	Cross sections of many gullies tend to be narrow relative to depth; sidewalls are steep; headcut usually prominent
Flow pattern develops as many small disconnected parallel channels ending at ephemeral cropland gullies, terrace channels, or where deposition occurs; they are generally uniformly spaced and sized	Usually forms a dendritic pattern along depressional water courses, beginning where overland flow, including rills, converge; flow patterns may be influenced by tillage, crop rows, terraces, or other unnatural features	Tend to form a dendritic pattern along natural water courses; nondendritic patterns may occur in road ditches, terrace, or diversion channels
Occurs on smooth side slopes above drainageways	Occurs along shallow drainageways upstream from incised channels or gullies	Generally occurs in well-defined drainageways
Soil is removed in shallow channels but annual tillage causes the soil profile to become thinner over the entire slope	Soil is removed along a narrow flow path, typically to the depth of the tillage layer where the untilled layer is resistant to erosion, or deeper where the untilled layer is less resistant; soil is moved into the voided area from adjacent land by mechanical action (tillage) and rill erosion, damaging an area wider than the eroded channel	Soil may be eroded to depth of the profile and can erode into soft bedrock

SOURCE: Foster et al. (in press).

offsite sedimentation analyses also require that ephemeral gully erosion be estimated because it can produce significant quantities of sediment.

No data on ephemeral gully erosion are available in the 1982 NRI. Furthermore, even though ephemeral gully erosion is related to some USLE factors, no attempt should be made to estimate ephemeral gully erosion from NRI data. Since future NRIs should take an inventory of this type of erosion, the remainder of this paper describes processes, physical impacts, policy implications, and inventory methods associated with ephemeral gully erosion to provide background to assist policymakers, NRI users, and USDA personnel.

EPHEMERAL GULLY EROSION PROCESSES

Ephemeral gully erosion is a process of flow detaching soil particles from a channel boundary and transporting the resulting sediment downstream (Foster, 1982a). A fundamental concept in erosion mechanics is that sediment load in flow is limited by either the transport capacity of the flow or the sediment available for transport, whichever is smaller (Foster, 1982b). Sediment available for transport in ephemeral gullies is from two sources--interrill and rill erosion on adjacent overland flow areas and erosion in upstream ephemeral gullies. The profile along many ephemeral gullies is concave, and grade decreases along them. Although transport capacity tends to increase as discharge increases along the gullies, the decrease in grade tends to lower transport capacity. The net result in gullies with a concave profile is that transport capacity increases to a maximum part way along and then decreases from there to the gully outlet (see Figure 1).

Sediment load increases along the channel from sediment added to the channel from adjacent overland flow areas and from sediment produced by upstream erosion in the channel. If channel grade decreases significantly, transport capacity equals sediment load somewhere downstream, at which point deposition begins and continues to the channel outlet, as Figure 1 illustrates. When the channel profile is only slightly concave, deposition may

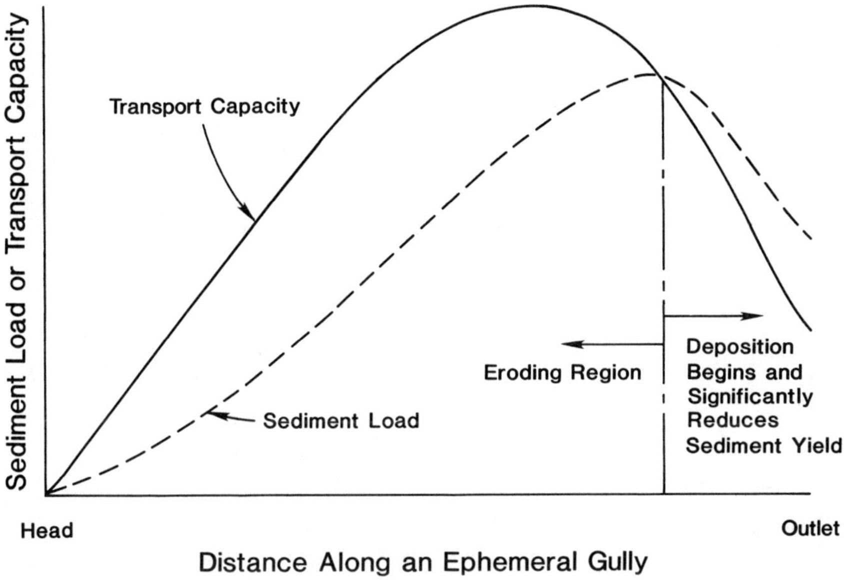

FIGURE 1 Variation of sediment load and transport capacity along a typical ephemeral gully having a concave profile that decreases in grade along the channel.

not occur. Backwater from a restricted channel outlet can also reduce transport capacity and cause deposition. When deposition occurs, sediment yield from the channel is largely controlled by the flow's transport capacity near the outlet of the channel rather than by the amount of upstream erosion (Foster, 1982a). When grade is uniform along a channel in a field, deposition may or may not occur, depending on the amount of runoff and sediment arriving from the adjacent overland flow areas. Deposition, if it occurs, is along the entire length of the channel, just as it occurs along uniform gradient terrace (Foster and Ferriera, 1981).

When channel profiles are convex, deposition can occur in an upper reach, where grade is relatively flat. Transport capacity increases along the channel faster than the sediment load does, causing deposition to end and erosion to begin in the channel (see Figure 2). The combination of steep grade and high flow rate near the outlet of a channel on a convex grade provides the potential for high erosion rates.

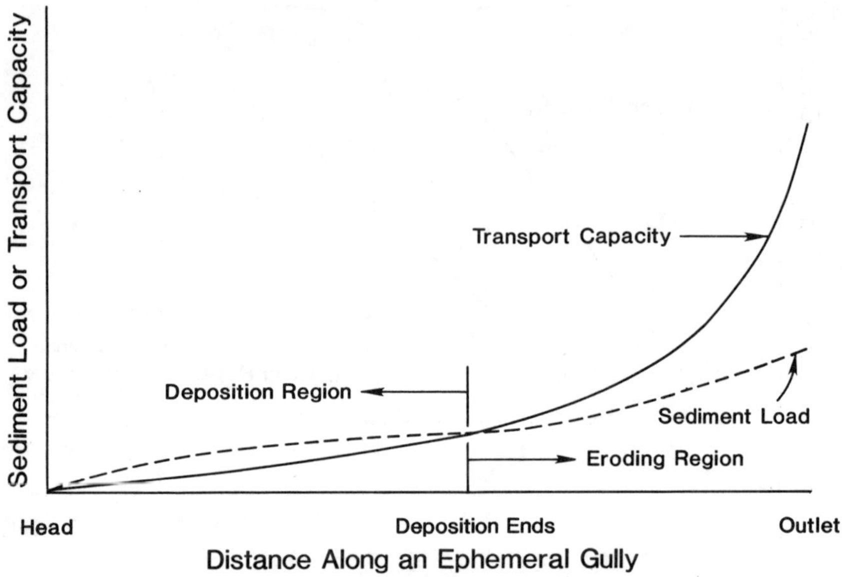

FIGURE 2 Variation of sediment load and transport capacity along a typical ephemeral gully having a convex profile that increases in grade along the channel.

Governing Equations

One erosion theory (Foster and Meyer, 1972a) holds that detachment rate depends on the fraction of the transport capacity filled by the sediment load, according to:

$$D_f = D_c(1 - G/T_c), \tag{1}$$

where D_f = detachment rate along the channel boundary [mass/(area • time)], D_c = detachment capacity of the flow [mass(area • time)], G = sediment load in the flow (mass/time), and T_c = transport capacity of the flow (mass/time).

Thus, maximum erosion occurs for a given flow rate when little sediment arrives from adjacent overland flow areas or from upstream erosion in the channel. A classic example of these conditions is stream degradation below a dam that has removed sediment from the flow (Knighton, 1984). As the sediment load fills the transport capacity (i.e., as the ratio G/T_c approaches 1), detachment rate is reduced, which appears to occur on some highly erodible soils in south Georgia. When G exceeds T_c, the term $1 - G/T_c$ becomes negative, indicating deposition. Thus, erosional, transport, and depositional processes in an ephemeral gully are directly related to runoff and sediment contributions from the overland flow areas of a field. Consequently, the erosion and sedimentation of an ephemeral gully cannot be evaluated without considering the field's hydrology and its rill and interrill erosion.

The mathematical relationship of these factors is given by the conservation of mass (continuity) equation (Foster and Meyer, 1972a):

$$dG/dx = D_l + D_f, \tag{2}$$

where x = distance down the channel, D_l = the contribution of sediment from adjacent overland flow areas, and D_f = detachment or deposition of sediment in the channel. This equation is for steady-state conditions. (Although thorough analysis of actual flow requires the more complex unsteady continuity equation [Bennett, 1974], Equation 2 suffices for many analyses, including NRI applications, and for discussion of erosion processes.) Integration of Equation 2 gives sediment load at any location along the channel as:

$$G = \int(D_l + D_f)dx, \tag{3}$$

where values for D_l could be from the USLE and values for D_f could be from Equation 1, which requires equations for D_c and T_c. A typical equation for detachment capacity (D_c) is (Ariathurai and Arulanandan, 1978; Foster and Lane, 1983):

$$D_c = K_c(\tau - \tau_c), \tag{4}$$

where K_c = a soil erodibility factor for erosion by flow, τ = shear stress of the flow acting on the channel boundary at a point in time and space, and τ_c = critical shear stress required to detach soil at a point in time and space.

Transport capacity (T_c) can be described by a similar equation, except that τ_c = critical shear stress required to move sediment after it has been detached and K_c = a transport factor K_t. Both K_t and τ_c for transport capacity are functions of particle diameter and density (Alonso et al., 1981).

Total erosion for a field during a storm is determined by integrating the equations in space over the channel network and over time as flow in the channel rises and falls. Between storms, cover and soil conditions change, affecting both the erosivity of the flow and the erodibility of the soil during later events. The equation for total detachment capacity during a storm at a location along a channel can be approximated by:

$$D_{ct} = \int K_c(\tau - \tau_c)dt, \qquad (5)$$

where D_{ct} = total detachment capacity at a location for the storm and t = time. Equation 5 can be integrated to give:

$$D_{ct} = K_c \int_{t_1}^{t_2} \tau dt - K_c \tau_c (t_2 - t_1), \qquad (6)$$

where t_1 = time when $\int$ exceeds $\int_c$, and t_2 = time when $\int$ falls below $\int_c$ (see Figure 3). In addition, Equation 6 can be approximated by:

$$D_{ct} = \beta K_c V\, AsC_c(1 - \tau_c/\beta A\sigma_p sC_c)^2, \qquad (7)$$

where β = a coefficient, V = runoff volume expressed as an average depth over the upstream drainage area, σ_p = peak runoff rate expressed as average depth over the drainage area per unit time, A = upstream drainage area drained by the location on an ephemeral gully, s = grade of the channel, and C_c = factor for cover conditions in the channel. The steps between Equations 6 and 7 are given in Appendix B.

Equation 7 identifies the major variables that should be considered in developing an empirical procedure to estimate ephemeral gully erosion. It represents potential sediment production at a particular ephemeral gully location during a storm. The total sediment that such erosion might produce in a field is determined by integrating Equation 7 along every branch of the gully network over the field, taking into account variations of A, s, and perhaps C_c and K_c along the channels. The value resulting from such an integration represents a maximum potential sediment production, which can then be reduced according to Equation 1 to account for sediment from rill and interrill erosion on adjacent overland flow areas.

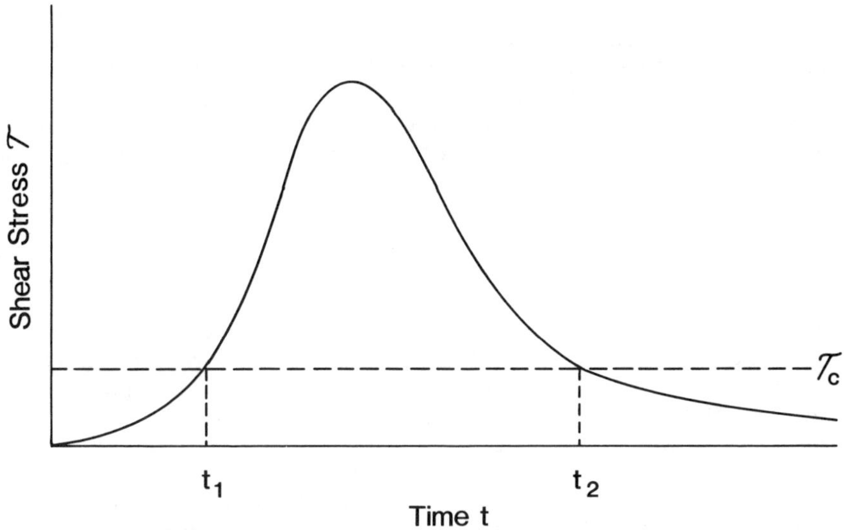

FIGURE 3 Variation of shear stress during a runoff event.

Factors Affecting Ephemeral Gully Erosion Runoff

The maximum flow rate must exceed a critical level with a given channel grade and cover if the shear stress of flow in an ephemeral gully is to exceed the critical shear stress of the soil. The parts of the drainage network where this does not occur, which varies within and from storm to storm, will experience no erosion. Flow rate in a channel, proportional to $A \cdot \sigma_p$, depends on rainstorm intensity and amount, infiltration characteristics of soil in the field, area and shape of the watershed, grade of the channel, and hydraulic roughness in the channel. The two most important rainstorm characteristics related to volume of runoff and peak runoff rate are storm depth and maximum intensity. The USLE erosivity term EI, E (storm energy) times I_{30} (maximum 30-minutes intensity), is a measure of these rainstorm characteristics (Foster et al., 1982c), which suggests that the erosivity factor of the USLE might be used as a climatic erosivity variable to estimate ephemeral gully erosion.

Runoff volume depends on a field's infiltration characteristics as affected by basic soil properties like soil texture and by management factors like cover and tillage. Hydrologic soil groupings and curve numbers used by the USDA Soil Conservation Service (SCS) indicate the runoff potential of a field (Knisel and Foster, 1981). An infiltration-based approach provides another method for estimating runoff (Brakensiek and Rawls, 1982).

Shear Stress of Flow

The shear stress that a flow exerts on a channel boundary is distributed between that acting on the soil and that acting on roughness elements like grass, crop residue, and clods (Foster et al., 1982b). The part acting on the soil is assumed to be responsible for detaching and transporting sediment. One reason that grassed waterways control ephemeral gully erosion is that grass significantly reduces the shear stress of the flow acting on the soil (Temple, 1980). The factor C_c in Equation 7 represents this effect and is approximately equal to the square of the ratio of flow velocity in a hydraulically rough channel to velocity in a smooth channel (Foster and Meyer, 1975). Similarly, crop residues left by conservation tillage reduce shear stress of the flow acting on the soil. As density of the cover increases, shear stress acting on the cover increases. If it exceeds the critical shear stress of the cover, the cover fails, and shear stress acting on the soil increases, which could cause serious erosion (Foster et al., 1982b).

Critical Shear Stress of Soil

Values for critical shear stress for soil (τ_c) have been a concern to channel designers for many years (ASCE, 1975). Generally, a channel is designed so that shear stress of the flow acting on the channel boundary is less than the critical shear stress, which provides for a stable, nonerodible channel (ASCE, 1975). Critical shear stress values have been related to a variety of soil properties, including soil texture, density, plasticity index, clay content, dispersion ratio, and sodium content (Ariathurai and Arulanandan, 1978; ASCE, 1975). Reported values vary greatly even for similar conditions, which suggests that critical shear stress values are difficult to define precisely.

Many earth-lined channels are constructed on consolidated soils, while natural channels in fields often occur on loose, tilled soil. Consequently, the effect of tillage, surface and buried residues, consolidation, plant roots, management, freezing and thawing, and other similar factors must be considered in any analysis of ephemeral gully erosion. On silt loam soils typical in the Midwest, tillage significantly decreases critical shear stress. A soil freshly tilled can be several times more erodible than one that has not been tilled for a year (Foster et al., 1982a). The rate that soil consolidates following tillage, and thereby increases critical shear stress, is not known, but limited experimental data suggest that significant increases can occur within 3 months (Foster et al., 1982d). This effect varies with soil; tillage on sandy soils may have less effect on critical shear stress than it does on soils high in clay content.

Nonerodible Layer and Previous Erosion

Ephemeral gully erosion in the Midwest is most obvious in the spring, when erosive rains occur on freshly prepared seedbeds. The surface-tilled soil has a low critical shear stress and is highly erodible. The underneath, untilled soil can have a high critical shear stress, be almost nonerodible, and act as a nonerodible layer. Flow quickly erodes through the tilled surface soil and stops at the untilled soil. With continued runoff, the channel widens and the erosion rate decreases (Foster and Lane, 1983). These channels are characteristically wide (6 to 12 feet) and shallow (4 to 8 inches deep). Subsequent storms that are smaller than the one that initially eroded the channel will cause little or no erosion. Conversely, had a small storm occurred first, much more erosion would occur subsequently (Foster, 1982a).

A factor F needs to be added to Equation 7 to represent this reduced potential for erosion because of previous erosion. Given a particular grade, critical shear stress, and hydraulic roughness for a channel, the final width of an eroded channel can be computed for a continuous, steady discharge rate (Foster and Lane, 1983). If the channel width from previous erosion is wider than the final width that the current storm could produce, no erosion occurs from the storm. But when the previous channel width is less than the potential final width for the current storm, the change in channel width can be approximated by (Foster and Lane, 1983):

$$\Delta W = [1 - \exp(-t_*)](W_f - W_i), \tag{8}$$

where ΔW = change in channel width, W_f = final channel width for the given discharge rate and channel conditions, and W_i = the channel width when the storm begins. The normalized time t_* is given by:

$$t_* = t(dW/dt)_i/(W_f - W_i), \tag{9}$$

where t = time, and $(dW/dt)_i$ = the initial rate at which channel is widening from its previous width. The factor F is proportional to the change in width (ΔW) in Equation 8.

When critical shear stress is uniform with depth, ephemeral gullies are incised, narrow, deep channels that have width-to-depth ratios that are much lower than those for ephemeral gullies on soil where the tilled surface is more erodible than the underneath, untilled soil (Foster and Lane, 1983). Field inspections of ephemeral gully erosion on loess soils in western Tennessee and northern Mississipi found channels that were much more incised than those in the Midwest. Whereas a single storm may cause most of the annual ephemeral gully erosion when the underneath, untilled soil acts as a nonerodible layer, each storm causes erosion in proportion to its erosivity on soils in the absence of a nonerodible layer. Therefore, the presence of a nonerodible layer greatly affects the distribution of ephemeral gully erosion over a year.

Probability of Erosive Event on Erodible Soil

A probability factor P also needs to be added to Equation 7 to account for the likelihood of an erosive storm occurring when a soil is highly susceptible to erosion. Thawing soil has a very low soil strength (Formanek, 1983) and can be very susceptible to erosion by flow (as is very obvious in the Palouse region). Ephemeral gully erosion can thus occur on no-till fields and in pastures, areas where the soil is consolidated and would normally be considered resistant to erosion. Soil thawing apparently reduces critical shear stress and leaves soils susceptible to erosion by runoff from rains occurring in late winter. In contrast, most ephemeral gully erosion on tilled land occurs in late spring immediately after secondary tillage for planting has left the soil susceptible to erosion. Over time, the soil consolidates following tillage and becomes much more resistant to erosion (Foster, 1982a).

Nonflow Detachment

Detachment of soil in an ephemeral gully can occur at times other than during a storm. Soil can slake and fall to the bottom of ephemeral gullies during nonflow periods, especially during the winter. The next major runoff cleans out this debris. Soil moisture in the channel banks can reduce soil strength, causing chunks of soil to slough into the channel (Piest et al., 1975a). Although a storm may be unable to detach soil, it may transport loose soil produced by these other detachment processes. Also, subsurface flow entering the channel can reduce a soil's critical shear stress, making the soil more erodible during a storm. Therefore, erosion may be greater in an ephemeral gully located on the landscape where subsurface flow exits the soil than it is in an ephemeral gully located on higher areas.

Headcut Advancement

Erosion is nonuniform at the upper end of those ephemeral gullies that are extended by the upstream advancement of a headcut or overfall. Local shear stress at the headcut can be very intense and cause locally intense erosion. Unfortunately, the mechanics of both flow and erosion at headcuts are not well understood. Uniform erosion is usually assumed in analyses of ephemeral gully erosion, which smooths erosion rates over some distance on either side of the headcut.

Deposition

Depositional areas, which are usually near the outlets of ephemeral gullies, can expand and contract during storms and from storm to storm. The location and amount of deposition depend on the flow's sediment load relative to its transport capacity (Foster and Huggins, 1977).

For example, when transport capacity decreases more slowly than does sediment load (before runoff ceases after rain ends during a storm), the location where deposition begins moves downstream, and the flow may erode previously deposited sediment. If a later storm occurs after significant canopy has developed over the field that reduces sediment production from rill and interrill erosion without lowering the transport capacity of flow in the ephemeral gully areas, the sediment available for transport in those gullies is reduced relative to transport capacity. The result is that the location where deposition begins moves downstream, and previously deposited sediment may be eroded.

Thus, since sediment deposited by previous storms can later be exposed to potentially erosive flows, the erodibility of deposited sediment must be considered. Deposited sediment that remains saturated during a storm is easily erodible (Foster et al., 1982d). Afterward, wetting and drying and other consolidating processes between runoff events can significantly increase the critical shear stress of deposited sediment, sometimes within 3 months (Foster et al., 1982d; Kemper et al., in press). Also, tillage mixes the deposited sediment with the underlying soil, making critical shear stress similar to that in other areas of the field.

Deposition usually occurs over a fairly broad area while erosion is an incisement process. Erosion removes soil from a smaller area than where deposition places the sediment. Thus, all previously deposited sediment may not be available to future eroding flows (Foster, 1982b).

Evolution of Landscape

The landscape is dynamic and evolves in response to erosion on it (Knighton, 1984). Ephemeral gully erosion occurs in the same locations each year and causes a drainage network to gradually become incised into the landscape. This incisement lowers the base level of adjacent overland flow slopes, which shortens overland flow slope lengths and steepens the landscape adjacent to the ephemeral gully areas. The increase in slope convexity and average steepness of adjacent overland areas may significantly increase rill and interrill erosion.

Control of Ephemeral Gully Erosion

Conservation tillage can satisfactorily control ephemeral erosion in less severe cases. In other situations, however, permanent channels like grassed waterways, terraces, and designed surface water disposal systems are needed. In the severest cases, additional permanent structures, such as concrete, rock, and corrugated metal structures that "drop" water to a lower elevation without causing erosion, may be needed to prevent an ephemeral gully from becoming a classical gully.

POLICY ISSUES ASSOCIATED WITH EPHEMERAL GULLY EROSION

Some of the national policy issues raised by ephemeral gully erosion are offsite sedimentation and water quality, onsite loss of productivity, inconvenience to farming operations, loss of land value, and quantification of benefits from treatment. Ephemeral gully erosion can represent a significant erosional area and sediment source within farm fields. Estimates made with the CREAMS (Chemicals, Runoff, and Erosion from Agricultural Management Systems) model (Knisel and Foster, 1981) and preliminary SCS measurements* suggest that sediment produced by ephemeral gully erosion can equal that produced by rill and interrill erosion. If it leaves fields, this sediment can cause more offsite sedimentation damage than would be expected by considering just rill and interrill erosion.

---

*SCS personnel are measuring and collecting field data on ephemeral gully erosion in about 30 states. The detail in these data varies greatly, but the information from Alabama, Georgia, and Maine is the most detailed. The SCS regional technical centers are assembling these data, and preliminary interpretations should be available in 1985. These data will be among the best that are available on ephemeral gully erosion. Potential users of the information should contact the SCS National Sedimentation Geologist (W. F. Mildner, National Sedimentation Geologist, USDA Soil Conservation Service, Washington, D.C., personal communication, 1984).

Sediment Yield

Consideration of offsite sedimentation and associated water quality issues must begin with knowing how much sediment actually leaves fields from and through ephemeral gullies. Less sediment may leave fields than is commonly assumed because deposition within fields may be greater than is currently estimated (Piest et al., 1975b). If, as expected, SCS field measurements continue to show that ephemeral gully erosion is producing considerable sediment, the delivery ratios (sediment yield/total erosion), now based on rill and interrill erosion alone, may require adjustment if estimates of ephemeral gully erosion are added directly to estimates of rill and interrill erosion.

Another issue concerns the use of fixed sediment delivery ratios for a given area as cover and management change. Such an assumption may be incorrect, and reduction in sediment yield from fields may not be proportional to reduction in either rill and interrill or ephemeral gully erosion. If transport capacity near the outlet of the field is controlling sediment yield, it must be reduced in order to lower sediment yield. Accurate estimates of sediment yield where ephemeral gully erosion is a major factor may require adjustments to current delivery ratio concepts and values. In fact, the simple but often used method of multiplying USLE estimates by a sediment delivery ratio is at best a very general way to estimate sediment yield; it needs to be improved.

Chemical Yield

Another offsite water quality issue associated with sediment is the concentrations of chemicals on the sediment yield. Such concentrations from ephemeral gully erosion are likely to be less than that from rill and interrill erosion because sediment from gullies is usually from deeper within the soil profile. This difference of concentration must be considered when chemical loss on sediment from a field is estimated. Although ephemeral gullies are a significant sediment source, channel reaches near their outlets can be major depositional areas, which reduces sediment yield from fields and enriches the sediment yield in fine particles. Since sediment-associated chemicals are carried by the fine particles, deposition enriches the concentration of chemicals on the sediment. Therefore, reduction of chemical yield is not proportional to reduction in sediment yield. Also, many agricultural pollutants may be soluble in the runoff and not associated with sediment (Knisel and Foster, 1981).

Crop Productivity

Onsite productivity issues must consider the loss of productivity within the eroded channel area and the loss of productivity on adjacent areas. Ephemeral gully erosion is very intense locally along its channels, which causes loss of the crop in the channel areas, but these areas are usually a small fraction of the total field (Thorne, 1984). Over the long term, incisement of ephemeral gullies steepens adjacent areas and accelerates rill and interrill erosion, and tillage drags soil into the eroded channels, further reducing soil depth and productivity on adjacent areas. The long-term productivity loss from ephemeral gully erosion extends, therefore, over an area larger than the immediate channels and over a long time.

The productivity issue must also consider whether a unit of ephemeral gully erosion averaged over a field has the same impact as a unit of rill and interrill erosion so averaged. Yet erosion and productivity loss over a field from rill and interrill erosion also vary. Estimates for a field are not accurate when based on an average erosion rate because of this variability of erosion and of nonlinearities in erosion/productivity relationships (Perrens et al., 1985).

Farming Operations

Few farmers allow ephemeral gullies to become classical gullies that divide fields, greatly reduce the efficiency of large farm equipment, and inconvenience farming operations. Nevertheless, over time the affected area grows as the landscape geomorphologically adjusts to accommodate the incised ephemeral gullies, which produces a variable and a less desirable and valuable landscape for farming. When ephemeral gully erosion is severe during a growing season, farmers must plow in the channels before harvest; when erosion is moderate, they plow in the channels before primary tillage to reduce wear-and-tear on equipment caused by crossing the eroded channels and to ensure uniform tillage in the vicinity of the ephemeral gullies.

Quantification of Benefits

Benefits of practices like terraces and grassed waterways and combinations of practices like terraces and conservation tillage in conservation systems include reduced ephemeral gully erosion. Historically, the reduction in rill and interrill erosion by soil conservation practices has been quantified with the USLE, and benefits from this reduction have been assigned. However, the reduction of erosion and associated benefits from practices like grassed waterways and other water disposal systems used to control ephemeral gully erosion have not been well quantified. To evaluate the total impact of erosion on farm fields, the amount of ephemeral gully erosion, the reduction in this erosion from installation of conservation practices, and the benefits from the reduction in this erosion must be estimated in addition to the common estimates of rill and interrill erosion. However, the technology required for these estimates does not exist but needs to be developed if ephemeral gully erosion is to be considered in public policy on erosion.

INVENTORY METHODS

Ephemeral gully erosion seems to have as much impact as rill and interrill erosion. Therefore, taking an inventory and analyzing this newly identified type of erosion is desirable for the 1987 NRI to determine all the damages caused by erosion and the full benefit of erosion control practices. Such information is needed to develop national policy on control of erosion on agricultural land.

If the 1987 NRI is to include an inventory of ephemeral gully erosion, ways to obtain the necessary field information must be chosen. The 1982 NRI contained no direct information on ephemeral gully erosion or sufficient information to estimate it. Before a method is applied, however, a sample area and the ephemeral gully network to be used at an NRI sample point must be identified.

Sample Area and Ephemeral Gully Network

Choosing an appropriate sample area over which to compute an average ephemeral gully erosion rate is a problem. Sampling along ephemeral gullies is normally limited to erosional areas because they can be readily identified and deposition has not been a major concern. A major spatial sampling question concerns the extent of any drainage network of ephemeral gullies that is to be sampled. For example, one branch of a field's network may experience significant erosion while an adjacent branch may experience none. The spatial average of ephemeral gully erosion will be higher if only the eroding branch and its drainage area are considered rather than the total watershed area drained by both branches. An area of a given size, perhaps 40 acres centered around an NRI sample point, could be used to inventory ephemeral gully erosion.

Another possible method for choosing the sample area is to trace the flow path from the sample point to the field outlet. The drainage area and the ephemeral gully network above this outlet point would be the sample area. Yet, field outlets may not be easily defined for lands other than cropland. Even if an accurate spatial average of ephemeral erosion can be obtained, the impact on productivity of variability of erosion over the sample area must be considered.

Once the specific sample area is determined, the drainage network of ephemeral gullies and their grades within the sample area must be established. Thorne (1984) has proposed an objective method to identify ephemeral gully areas based on 2-foot interval contour maps, convexity of the contours, upslope contributing area, and local slope gradient. This method does not require evidence of ephemeral gully erosion to identify the drainage network. Also, it provides overland flow slope lengths, which would be helpful in USLE applications as choice of USLE slope length continues to be inconsistent. As an alternative, the ephemeral gully network could be mapped in the field or drawn from interpretation of aerial photographs (Frazier et al., 1983). But this requires visual evidence of ephemeral gully erosion, which may not be present when the site is visited or photographed.

When the sample area and drainage network have been determined, a way to estimate the erosion in the channels must be selected and applied. Estimates of ephemeral

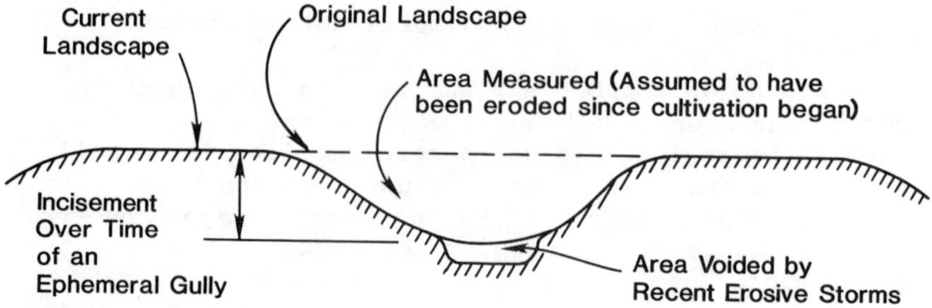

FIGURE 4 Estimating ephemeral gully erosion by measuring volume of landscape assumed to have been voided by ephemeral erosion since cultivation began.

gully erosion can be obtained in one of three ways--direct field measurement, estimation with equations, or a combination of the two.

Field Measurement

The SCS is using one of two methods to collect field data on ephemeral gully erosion. One method measures voided cross sections and reach lengths along the ephemeral gully network following erosive events. A difficulty with this approach is ensuring that the sample represents average ephemeral gully erosion over the field and average annual erosion. These measurements should be made over several years to establish averages. Also, accelerated erosion on areas adjacent to the ephemeral gullies is not measured with this method.

The second method overcomes this problem by sampling across the landscape (see Figure 4). This procedure assumes that both the time that the sample area has been cultivated and the original landscape when cultivation began are known. It directly gives an average annual estimate without having to consider the representativeness of particular erosive events. This method is being used by SCS in Alabama on land that has been in cultivation for about 30 years and has experienced severe ephemeral gully erosion.

Measurements with both methods must be taken over the ephemeral gully network to obtain a field average. An alternative to field surveys is to use stereographic photography, being developed by scientists with the University of Georgia, Washington State University, and

USDA Agricultural Research Service (ARS) at Watkinsville, Georgia; Pullman, Washington; and Treynor, Iowa (Frazier et al., 1983; Spomer and Mahurin, 1984; Welch et al., 1984).

Mathematical Prediction

Three types of mathematical procedures could be used to estimate ephemeral gully erosion. One is an empirical factor approach (similar to Equation 7) being developed by Thorne (1984) and scientists at the ARS Sedimentation Laboratory in Oxford, Mississippi. The second mathematical approach is the use of theoretically based equations like those being developed by Iowa State University scientists and ARS scientists at Iowa State University in Ames, Iowa. The third method is a simulation approach that uses fundamental concepts and equations like those being developed by University of Kentucky scientists and ARS scientists at Tucson, Arizona; Fort Collins, Colorado; and West Lafayette, Indiana (Foster and Lane, 1983; Foster et al., 1983; Hirschi and Barfield, 1984).

Thorne's (1984) preliminary empirical equation is given by:

$$E = \alpha[F_f K_f(\phi - \phi_c)]\ C_f, \qquad (10)$$

where E = ephemeral gully erosion, α = a coefficient, F_f = flow erosivity factor, K_f = soil erodibility factor for flow, ϕ = an index that defines areas susceptible to ephemeral gully erosion ($\phi = A\zeta/s$, A = upstream area, s = channel grade, and ζ = contour convexity), ϕ_c = a critical value for ϕ (no ephemeral erosion when $\phi < \phi_c$), and C_f = a cover-management factor. Equation 10 will be fitted to the data being collected by SCS to determine parameter values and to validate the method.

The theoretically based equations will likely involve some combination of Equations 1 through 9 plus other equations that consider evolution of eroded channel shapes (Foster and Lane, 1983). Perhaps they will use the USLE erosivity index to describe the erosivity of climate. Parameters from the drainage network will include degree of concavity or convexity of channel profiles, average channel grade, degree of branching, and length of the network branches. Data being collected by scientists in field and laboratory experiments at several locations will be used to determine many of the parameter values for this method (e.g., Foster et al., 1982d; Hirschi and Barfield, 1984). The field data being collected by SCS will also be used to validate the method and determine some parameter values.

Current hydrology-erosion simulation models like ANSWERS, CSU, CREAMS1, and CREAMS2 (Beasley et al., 1980; Foster et al., 1983; Knisel and Foster, 1981; Simons et al., 1975) can also be used to estimate ephemeral gully erosion. These models require a great deal of input data and computers to drive them and do not seem practical for present NRI applications. Their major applications are evaluation and planning at specific sites. Also, more research is needed to determine their parameter values over a wide range of field conditions. However, data from research being conducted to develop the other methods can be used to develop and validate the simulation models.

Empirical prediction methods will probably be available by 1987, while a more fundamental method will be available by 1990 for use in NRIs. If necessary, field monitoring could be used to collect NRI data on ephemeral gully erosion, but it could be expensive. The method used in the 1987 NRI will likely be a combination of a field survey and an empirical factor method.

Cautions

Before prediction methods for ephemeral gully erosion become available, watershed planners and others are anxious to develop and use estimates from the SCS field data. The amount of detail available ranges from little to extensive. Some users are satisfied with very general figures--for example, that ephemeral gully erosion is about two-thirds of rill and interrill erosion in most fields. Although such statements may be generally true in a particular area like Alabama, they can be grossly wrong in other parts of the country because of major differences in climate, soil, cover, and management. Great care should be used, therefore, when transferring simple relationships derived from measured data on ephemeral gully erosion from one part of the United States to another and even from one soil or cropping practice to another. Also, given the difficulties with representative sampling in space and time, users of a particular data set should verify that the data were not biased toward the more severe cases.

If ephemeral gully erosion could be related to rill and interill erosion by multiplying USLE estimates by a simple factor, as Osborn et al. (1977) suggested, it could be readily estimated. Although some relation between these types of erosion must exist, it is not reliable in many cases. The USLE is a lumped equation representing erosion processes of detachment by rainfall, detachment by flow, transport by flow, and deposition in microareas. Thus, it includes many other factors besides those important in ephemeral gully erosion. For example, the USLE cover factor, a lumped parameter, underestimates the effect of cover on detachment by flow because cover reduces rill erosion more than it does interrill erosion (Hussein and Laflen, 1982). Also, the USLE does not factor in critical shear stress, which is more important in ephemeral gully erosion than in rill and interrill erosion. The result is that ephemeral gully erosion can be slight in a field where rill and interrill erosion is great. Furthermore, USLE slope length and steepness factors may not be highly correlated with features of an ephemeral gully network. Therefore, unless data become available that show otherwise, multiplication of USLE estimates by a factor to estimate ephemeral gully erosion is not recommended.

SUMMARY

Topography often causes overland flow to collect in a few major natural waterways before leaving fields. These waterways are concentrated flow areas, and profiles along them are often concave, resulting in erosion in upper reaches and deposition in lower reaches. These gully-like areas are short-lived--hence the term ephemeral--because they are annually plowed in during farming. Unlike rills, these eroded channels are reformed each year in the same locations and gradually become incised in the landscape, a process that steepens adjacent overland flow slopes and accelerates rill and interrill erosion on them. Thus the impact of ephemeral gully erosion extends over a significantly larger field area than just the immediate eroded channel area.

The basic equation often used to describe this process is that erosion rate is proportional to the difference between the shear stress of flow in the ephemeral gully and the soil's critical shear stress. Flow's shear stress is related, in turn, to the channel's flow rate, grade, and cover. Of course, flow from runoff is related to storm characteristics, infiltration (as that is affected by soil, cover, and management in the field), and watershed shape and area. The critical shear stress of the soil varies with soil properties, especially as they are modified by climate, tillage, and management. Tillage leaves some soils highly susceptible to erosion by flow. Ephemeral gullies on freshly tilled soils are often restricted by the underlying untilled soil, and the eroded channels tend to be wide and shallow. Channels on soils where no layer restricts downward erosion, on the other hand, tend to be narrow and incised.

Ephemeral gully erosion is highly variable in space and time, which makes sampling for field measurements difficult and estimated erosion rates subject to large errors. Ephemeral gully erosion was not estimated in the 1982 NRI, but its importance suggests that it should be estimated in the 1987 NRI. Inventories will probably be conducted by making field measurements for data to be put into an empirical prediction method. Multiplying USLE rill and interrill erosion estimates by an ephemeral gully erosion factor seems inappropriate. Likewise, conclusions based on field measurements for one region, soil, and management practice may not be transferable to other conditions.

Ephemeral gully erosion can lower productivity over a significant portion of many fields on the areas adjacent to the ephemeral gullies. It produces a quantity of sediment that approaches that from rill and interrill erosion in many fields, an important consideration in analyses of offsite impacts from sediment. The channels associated with this newly identified erosion are the main delivery system for water and sediment from most fields, and thus deposition common in these channels must be considered in offsite impact analyses. Although most farmers do not allow these channels to grow into gullies too large to cross with farm equipment, ephemeral gully erosion can inconvenience farming operations. Its long-term reshaping of the landscape can reduce land value. If these impacts could be more clearly identified, the benefits of control practices--including conservation tillage, grassed waterways, terraces, and other water disposal systems--could be better established, which is important for establishing a national soil conservation policy.

REFERENCES

Alonso, C. V., W. H. Neibling, and G. R. Foster. 1981. Estimating sediment transport capacity in watershed modeling. Trans. ASAE 24:1211-1220.

Ariathurai, R., and K. Arulanandan. 1978. Erosion rates of cohesive soils. J. Hydraulics Div., Proc. Am. Soc. Civil Engineers 104(HY2):279-283.

ASCE (American Society of Civil Engineers). 1975. Sedimentation Engineering. New York: American Society of Civil Engineers.

Beasley, D. B., L. F. Huggins, and E. J. Monke. 1980. ANSWERS: A model for watershed planning. Trans. ASAE 23:938-944.

Bennett, J. P. 1974. Concepts of mathematical modeling of sediment yield. Water Resourc. Res. 10:485-492.

Brakensiek, D. L., and W. J. Rawls. 1982. An infiltration based rainfall-runoff model for SCS Type 2 distribution. Trans. ASAE 25:1607-1611.

Formanek, G. E. 1983. Shear Strength of a Thawing Silt Loam Soil: An Approach to Erodibility. M.S. thesis. Washington State University, Pullman.

Foster, G. R. 1982a. Channel erosion within farm fields. Preprint 82-007. American Society of Civil Engineers. New York.

Foster, G. R. 1982b. Modeling the soil erosion process. Pp. 297-382 in Hydrologic Modeling of Small Watersheds, C. T. Hann, H. P. Johnson, and D. L. Brakensiek, eds. St. Joseph, Mich.: American Society of Agricultural Engineers.

Foster, G. R. 1982c. Relation of USLE factors to erosion on rangeland. Pp. 17-35 in Proc. of Workshop on Estimating Soil Erosion and Sediment Yield from Rangelands. ARM-W-26. Washington, D.C.: USDA Agricultural Research Service.

Foster, G. R., and V. A. Ferreira. 1981. Deposition in uniform grade terrace channels. Pp. 185-197 in Crop Production with Conservation in the '80s. Proc. Am. Soc. Agric. Eng., December 1-2, 1980. Publ. 7-81. St. Joseph, Mich.: American Society of Agricultural Engineers.

Foster, G. R., and L. F. Huggins. 1977. Deposition of sediment by overland flow on concave slopes. Pp. 167-182 in Soil Erosion: Prediction and Control. Special Publ. No. 21. Ankeny, Iowa: Soil Conservation Society of America.

Foster, G. R., and L. J. Lane. 1983. Erosion by concentrated flow in farm fields. Pp. 9.65-9.82 in Proc. of the D. B. Simons Symposium on Erosion and Sedimentation. Ft. Collins: Colorado State University.

Foster, G. R., and L. D. Meyer. 1972a. A closed-form soil erosion equation for upland areas. Chapter 12 in Sedimentation (Einstein), H. W. Shen, ed. Fort Collins: Colorado State University.

Foster, G. R., and L. D. Meyer. 1972b. Transport of soil particles by shallow flow. Trans. ASAE 15:99-102.

Foster, G. R., and L. D. Meyer. 1975. Mathematical simulation of upland erosion by fundamental erosion mechanics. Pp. 190-207 in Present and Prospective Technology for Predicting Sediment Yields and Sources. ARS-S-40. Washington, D.C.: USDA Science and Education Administration.

Foster, G. R., C. B. Johnson, and W. C. Moldenhauer. 1982a. Critical slope lengths for unanchored cornstalk and wheat straw residue. Trans. ASAE 25:935-939.

Foster, G. R., C. B. Johnson, and W. C. Moldenhauer. 1982b. Hydraulics of failure of unanchored cornstalk mulches for erosion control. Trans. ASAE 25:935-939, 947.

Foster, G. R., F. Lombardi, and W. C. Moldenhauer. 1982c. Evaluation of rainfall-runoff erosivity factors for individual storms. Trans. ASAE 25:124-129.

Foster, G. R., W. R. Osterkamp, L. J. Lane, and D. W. Hunt. 1982d. Effect of discharge on rill erosion. Paper No. 82-2572. St. Joseph, Mich.: American Society of Agricultural Engineers.

Foster, G. R., R. E. Smith, W. G. Knisel, and T. E. Hakonson. 1983. Modeling the effectiveness of onsite sediment controls. Paper No. 83-2092. St. Joseph, Mich.: American Society of Agricultural Engineers.

Foster, G. R., L. J. Lane, and W. F. Mildner. In press. Seasonally ephemeral cropland gully erosion. In Proc. of the ARS-SCS Natural Resources Modeling Workshop. Washington, D.C.: USDA Agricultural Research Service.

Frazier, B. E., D. K. McCool, and E. F. Engle. 1983. Soil Erosion in the Palouse: An aerial perspective. J. Soil Water Conserv. 38:70-74.

Hirschi, M. C., and B. J. Barfield. 1984. Modeling channel erosion with emphasis on upland areas. Paper No. 84-2549. St. Joseph, Mich.: American Society of Agricultural Engineers.

Hussein, M. H., and J. M. Laflen. 1982. Effects of crop canopy and residue on rill and interrill soil erosion. Trans. ASAE 25:1310-1315.

Hutchinson, D. E., and H. W. Pritchard (for committee). 1976. Resource conservation glossary. J. Soil Water Conserv. 31:1-63.

Kemper, W. D., T. J. Trout, M. J. Brown, D. L. Carter, and R. C. Rosenau. In press. Factors affecting furrow erosion. In Proc. of the ARS-SCS Natural Resources Modeling Workshop. Washington, D.C.: USDA Agricultural Research Service.

Knighton, D. 1984. Fluvial Forms and Processes. London: Edward Arnold.

Knisel, W. G., and G. R. Foster. 1981. CREAMS: A system for evaluating best management practices. Pp. 177-194 in Economics, Ethics, Ecology: Roots of Productive Conservation. Ankeny, Iowa: Soil Conservation Society of America.

Meyer, L. D. 1981. How rain intensity affects interrill erosion. Trans. ASAE 24:1472-1475.

Moss, A. J., P. H. Walker, and J. Hutka. 1979. Raindrop-stimulated transportation in shallow water flows: An experimental study. Sedimentary Geology 22(3-4):165-184.

Osborn, H. B., J. R. Simanton, and K. G. Renard. 1977. Use of the Universal Soil Loss Equation in the semiarid Southwest. Pp. 41-49 in Soil Erosion: Prediction and Control. Special Publ. No. 21. Ankeny, Iowa: Soil Conservation Society of America.

Perrens, S. J., G. R. Foster, and D. B. Beasley. 1985. Erosion's effect on productivity along nonuniform slopes. Pp. 201-214 in Erosion and Soil Productivity. Publ. No. 8-85. St. Joseph, Mich.: American Society of Agricultural Engineers.

Piest, R. F., J. M. Bradford, and R. G. Spomer. 1975a. Mechanisms of erosion and sediment movement from gullies. Pp. 162-176 in Present and Prospective Technology for Predicting Sediment Yields and Sources. ARS-S-40. Washington, D.C.: USDA Science and Education Administration.

Piest, R. F., L. A. Kramer, and H. G. Heinemann. 1975b. Sediment movement from loessial watersheds. Pp. 130-141 in Present and Prospective Technology for Predicting Sediment Yields and Sources. ARS-S-40. Washington, D.C.: USDA Science and Education Administration.

Schertz, D. L. 1983. The basis for soil loss tolerances. J. Soil Water Conserv. 38:10-14.

Simons, D. B., R. M. Li, and V. A. Stevens. 1975. Development of Models for Predicting Water and Sediment Routing and Yield from Storms on Small Watersheds. Report CER74-75 DBS-RML-VAS 24. Ft. Collins: Colorado State University.

Spomer, R. G., and R. L. Mahurin. 1984. Time-lapse remote sensing for rapid measurement of changing landforms. J. Soil Water Conserv. 39:397-401.

Temple, D. M. 1980. Tractive force design of vegetated channels. Trans. ASAE 23:884-890.

Thorne, C. R. 1984. Prediction of Soil Loss Due to Ephemeral Gullies in Arable Fields. Report CER83-84CRT. Ft. Collins: Colorado State University.

Welch, R., T. R. Jordan, and A. W. Thomas. 1984. A photogrammetric technique for measuring soil erosion. J. Soil Water Conserv. 39:191-194.

Wischmeier, W. H., and D. D. Smith. 1978. Predicting Rainfall Erosion Losses: A Guide to Conservation Planning. Agriculture Handbook No. 537, USDA Science and Education Administration. Washington, D.C.: U.S. Government Printing Office.

APPENDIX A: APPLICATION OF THE USLE IN THE 1982 NRI: SLOPE LENGTH AND STEEPNESS FACTORS

The 1982 NRI used the Universal Soil Loss Equation to estimate sheet and rill erosion at a sample point. The instructions on the worksheet for recording the field data for slope length and steepness were:

> Enter the length of slope in feet through the point. On terraced land, enter the distance between terraces. Slope length is the distance from the point of origin [whether on or off the PSU (primary sampling units)] of overland flow to either of the following: i) the point where the slope decreases to the extent that deposition of sediment begins, or ii) the point where runoff enters an area of concentrated flow or a channel. Enter the percent slope to the nearest percent on slopes greater than one; enter to the nearest 0.1 percent for slopes less than one. Do not enter "0." Measure slope percent on the segment of landform on which the point falls. Measure in the direction that water would flow overland.

Using these slope length and steepness values, what does the calculated erosion represent? To illustrate, erosion rates were computed for segments along a typical complex-shaped land profile that varies from 2 percent to 8 percent to 3 percent steepness (see Table A-1). The computations were made using the 1982 NRI procedure and a procedure specifically designed to compute the average soil loss for a slope segment (Foster and Wischmeier, 1974; Renard and Foster, 1983; Wischmeier and Smith, 1978).

The apparent intent of the NRI procedure was to provide an estimate of average erosion over the landscape when the erosion rates at many sample points are averaged. A uniform distribution of the sample points over the landscape was assumed. Even when this assumption is met, however, the 1982 NRI method is only correct for uniform land profiles. The error in the method for computing average soil loss for an irregular profile depends on the degree of curvature of the profile--the greater the curvature, the greater the error. In the example shown

TABLE A-1 Erosion Rates Along a Nonuniform Slope

j	λ_j	s_j	S_j	m_j	ω_j	$(\lambda/\lambda_u)^{m_j}$	Col. 4x6x7	Col.8 xk	Irreg. A_{jI}	NRI A_{jN}	A_{cj}
(1)	(2)	(3)	(4)	(5)	(6)	(7)	(8)	(9)	(10)	(11)	(12)
1	40	2	0.18	0.3	0.12	1.36	0.03	0.15	2.0	3.2	3.2
2	80	4	0.35	0.4	0.17	1.50	0.09	0.46	5.8	6.8	6.8
3	120	8	0.84	0.5	0.21	1.66	0.30	1.49	19.0	18.0	18.0
4	160	6	0.57	0.5	0.25	1.66	0.24	1.19	15.3	12.2	12.2
5	200	3	0.26	0.3	0.25	1.36	0.09	0.45	5.7	4.5	4.5

Key:
(1) Segment index.
(2) Distance to lower end of segment (feet).
(3) Slope steepness of segment (percent).
(4) USLE slope factor value for segment.
(5) USLE slope length exponent for segment.
(6) $\omega_j = (j/k)^{m+1} - [(j-1)/k]^{m+1}$, where k = number of slope segments (5).
(7) λ = total slope length (200 feet); λ_u = unit plot length (72.6 feet).
(8) Product of columns 4, 6, and 7, $S_j\omega_j(\lambda/\lambda_u)^{m_j}$; summation of column = $\overline{LS}$ to compute average soil loss for the entire slope as RKLSCP; $\bar{A}$ = 8.9.
(9) Product of column 8 and k, gives $(LS)_j$ value to compute average soil loss for the segment.
(10) Average soil loss (tons/acre) for the segment as computed by the irregular slope procedure (Wischmeier and Smith, 1978); A_{jI} = RKCP x column 8, where R = 100 EI units, K = 0.32 tons/(acre • EI unit), C = 0.4, and P = 1.0 in this example; $\overline{LS}$ = 0.75.
(11) Average soil loss (tons/acre) for the segment as computed by NRI method; $A_{jN} = RKCP \times S_j \times (\lambda/\lambda_u)^{m_j}$; $\bar{A}$ = 9.6.
(12) Soil loss to compare with soil loss tolerance; $A_{cj} = A_{jI}/(k\omega_j)$.

in Table A-1, the NRI method underestimates average soil loss for the profile by 7 percent, which is not great considering other errors in USLE estimates. However, this error is systematic, whereas other errors would be random. The 7 percent error, if corrected, would change a soil loss of 5.0 tons/acre to 5.4 tons/acre.

The errors in average soil loss for individual segments are greater than for average soil loss for the profile. For example, average soil loss by the NRI method for the first slope segment in Table A-1 is 3.2 tons/acre versus the correct value of 2.0 tons/acre, an error of 60 percent. On the last segment, the NRI estimate is 4.5 tons/acre versus the correct 5.7 tons/acre, an error of 21 percent. Thus, the NRI method overestimates soil loss for sample points at the top of the slope and underestimates it near the end of the slope. These errors will be apparent and significant when the data are summarized according to a classification that divides slope lengths on a landscape.

To estimate soil loss at a point (Renard and Foster, 1983), the USLE is applied as:

$$A_j = (1 + m_j)RK_j(\lambda_j/\lambda_u)^{m_j}S_jC_jP_j, \quad \text{(A-1)}$$

where A_j = soil loss at point j, m_j = USLE slope length exponent for the slope steepness at point j, λ_j = slope length to point j, λ_u = length of the unit plot (72.6 feet), R = rainfall erosivity factor, and K_j, S_j, C_j, and P_j are USLE factor values at point j for soil erodibility, steepness, cover-management, and supporting practices factors, respectively.

According to Equation A-1, the soil loss at the lower end of a uniform slope is 1 + m (1 + m = 1.5 for slopes steeper than 5 percent since m = 0.5) times the average soil loss for the entire uniform slope, which means that over 60 percent of a uniform slope is eroding at a rate in excess of the soil loss tolerance value when the average soil loss for the slope (the value normally computed with the USLE) equals the soil loss tolerance value. Furthermore, the calculated soil loss over the last 20 percent of a uniform slope is 40 percent in excess of soil loss tolerance when the average soil loss equals soil loss tolerance. This range of soil loss variation along a uniform slope is usually neglected because of imprecision in soil loss tolerance values. However, application of the USLE irregular slope procedure (Wischmeier and Smith, 1978) requires soil loss values for individual slope segments that are on an equal basis for comparison to soil loss tolerance values. That adjustment can be made with the equation:

$$A_{cj} = A_j/(k\omega_j), \quad \text{(A-2)}$$

where A_{cj} = soil loss for a slope segment to compare with the soil loss tolerance, A_j = average soil loss for a slope segment, k = number of slope segment, ω_j = $(j/k)^{m+1} - [(j - 1)/k]^{m+1}$, and j = slope segment index.

This computation of soil loss removes the effect of the position on the land profile. Note from Table A-1 that values for A_{cj} (the last column) equal those computed with the NRI method. The values from the NRI method and from Equation A-2 are average soil loss values for a slope λ long of steepness s_j, and they can be compared directly with soil loss tolerance values.

Erosion, as column 10 in Table A-1 shows, varies greatly along a slope. Preferably, future NRIs would compute soil loss at a point according to Equation A-1. This soil loss value would be compared with a soil loss tolerance value to determine if erosion is a problem at the sample point. If such a procedure is followed, present soil loss tolerance values need adjustment to reflect permissible soil loss at a point rather than average soil loss over a uniform slope, as they now do. As research on the impact of erosion on productivity progresses, new soil loss tolerance concepts should recognize variation of soil loss over the landscape and define soil loss tolerance values that can be applied at a point on the landscape or at least to a slope segment that is as short as one-fifth of the slope length. Use of average soil loss for a slope length can seriously underestimate the impact of erosion on productivity on slopes where erosion varies greatly (Perrens et al., 1985).

REFERENCES

Foster, G. R., and W. H. Wischmeier. 1974. Evaluating irregular slopes for soil loss prediction. Trans. ASAE 17:305-309.

Renard, K. G., and G. R. Foster. 1983. Soil conservation: Principles of erosion by water. Pp. 155-176 in Dryland Agriculture. Agronomy Monograph No. 23. Madison, Wis.: American Society of Agronomy.

Wischmeier, W. H., and D. D. Smith. 1978. Predicting Rainfall Erosion Losses: A Guide to Conservation Planning. Agriculture Handbook No. 537, USDA Science and Education Administration. Washington, D.C.: U.S. Government Printing Office.

APPENDIX B: DERIVATION OF EROSION EQUATION FOR EPHEMERAL GULLY EROSION

The basic governing equation for the capacity of flow to detach soil at a cross section along an ephemeral gully is:

$$D_{ct} = K_c \int_{t_1}^{t_2} \tau dt - K_c \tau_c (t_2 - t_1), \qquad \text{(B-1)}$$

where D_{ct} = total detachment capacity for a storm, K_c = a soil erodibility factor for detachment by flow, τ = flow shear stress, τ_c = critical shear stress of the soil, t = time, t_1 = time that exceeds τ_c, and t_2 = time that τ becomes less than τ_c (see Figure 3).

Usually the function τ versus t is too complex to integrate analytically, and the function can vary greatly from storm to storm. Simulation models like CREAMS2 (Foster et al., 1983) numerically generate and integrate the τ versus t function. Many planning and inventory operations can use an empirical and approximate approach, which leads to the proposed equation of:

$$D_{ct} = \beta K_c V A s C_c \,(1 - \tau_c/\beta A \sigma_p s C_c)^2, \qquad \text{(B-2)}$$

where D_{ct} = total detachment capacity for the storm, β = a coefficient, K_c = a soil erodibility factor for detachment by flow, V = runoff volume expressed as an average depth over the upstream drainage area, A = upstream drainage area drained by the location on an ephemeral gully, s = grade of the channel, C_c = factor for cover conditions in the channel, τ_c = critical shear stress of the soil, and σ_p = peak runoff rate expressed as average depth over the drainage area per unit time. The purpose of this appendix is to derive this approximate equation.

For simplicity, the τ versus t function can be rearranged and approximated as shown in Figure B-1. Shear stress τ varies with time as αt and the integral $\int \tau dt$ is:

$$\int_{t_1}^{t_2} \tau dt = \alpha(t_2^2 - t_1^2)/2, \quad \text{(B-3)}$$

which can be factored to give:

$$\alpha(t_2^2 - t_1^2)/2 = \alpha(t_2 - t_1)(t_2 + t_1)/2. \quad \text{(B-4)}$$

Since $\alpha t_2 = \tau_p$ and $t_1 = \tau_c$:

$$\alpha(t_2 - t_1)(t_2 = t_1)/2 = (t_2 - t_1)(\tau_p + \tau_c)/2. \quad \text{(B-5)}$$

Substituting Equation B-5 in Equation B-1 gives:

$$D_{ct} = K_c(t_2 - t_1)\tau_p(1 - \tau_c/\tau_p)/2. \quad \text{(B-6)}$$

An approximation of t_p is (Foster et al., 1982b):

$$\tau_p = \beta Q_p s C_c, \quad \text{(B-7)}$$

where β = a coefficient, and Q_p = peak discharge rate. However, Q_p can be approximated by:

$$Q_p = A\sigma_p. \quad \text{(B-8)}$$

Time t_2 is the duration of the runoff and can be approximated by:

$$t_2 = 2V/\sigma_p. \quad \text{(B-9)}$$

The time $t_2 - t_1$ can be approximated from the proportionalities of the triangle in Figure B-1 as:

$$(t_2 - t_1)/t_2 = (\tau_p - \tau_c)/\tau_p \quad \text{(B-10)}$$

or:

$$t_2 - t_1 = t_2(1 - \tau_c/\tau_p). \quad \text{(B-11)}$$

The substitution of Equations B-7, B-9, and B-11 in Equation B-6 yields Equation B-12:

$$D_{ct} = K_c\ (V/\sigma_p)(\beta A\sigma_p s C_c)(1 - \tau_c/\ \beta A\sigma_p s C_c)^2, \quad \text{(B-12)}$$

which reduces to Equation B-2.

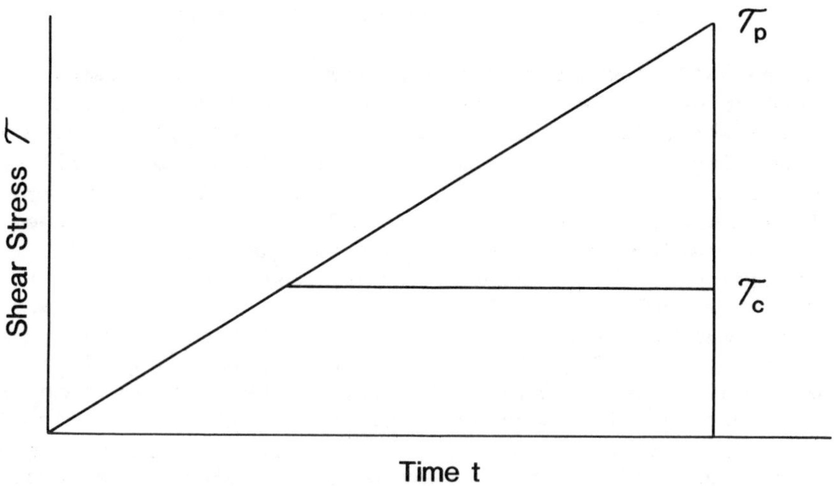

FIGURE B-1 Approximation of the shear stress τ versus time t function.

Clearly, Equation B-2 is very approximate, but it illustrates the important variables in estimating ephemeral gully erosion and a possible arrangement of terms in an empirical equation.

Discussion

B. J. Barfield and J. C. McBurnie

Foster's review of ephemeral gully erosion provides an excellent overview of the present state of our understanding of the physical processes involved in the movement of soil in the channelized flow areas. In addition, his discussion of the effects of ephemeral gullies on sediment yield, crop yield, and chemical content of runoff adequately describes our understanding of these processes.

In this response, some of Foster's points will be restated for emphasis. Additionally, remarks will be given relating some of the serious limitations of present modeling efforts.

EPHEMERAL GULLY EROSION AND THE NRI

As discussed in Foster's review, ephemeral gullies tend to be stable landscape features that form in those areas where flow is concentrated into significant channels resulting from nonuniformities in the landscape. Rill erosion, on the other hand, occurs in small nonpermanent flow channels normally spread randomly over a hillslope. An independent estimate of ephemeral gully erosion is necessary since the Universal Soil Loss Equation (USLE), which is the major tool for predicting soil erosion, was developed from a data base that did not include channelized flow. As Foster points out, ephemeral gully erosion can be a large percentage of the total erosion on a watershed. Although not estimated in the 1982 National Resources Inventory (NRI), it is likely that erosion from ephemeral gullies will be estimated in future inventories. An understanding of the processes is imperative.

FIELD MEASUREMENT OF EPHEMERAL GULLIES

Foster presents an excellent review of techniques for measuring ephemeral gully erosion. Several cautions seem appropriate in connection with the methods:

- Ephemeral gully erosion is likely to be highly variable, depending on inherent geomorphic characteristics such as soils, landscape relief, slope, cover, cultural methods of an individual farmer, and climatic variability.
- Changes in prevailing cultural practices are likely to make estimates from samples across the landscape unreliable predictors of current erosion rates.
- Changes in susceptibility to erosion with crop stage and the stochastic variability of erosive precipitation necessitate the collection of erosion data from many storms over several years to develop reliable estimates. Data on the time distribution of erosive storm and cover should be collected during the sampling period and compared to long-term averages.
- Projections of data from one climatic region to another is not advisable. Quite possibly, projection from one watershed to another may lead to erroneous results.

MATHEMATICAL PREDICTION MODELS

Foster's review includes a detailed discussion of equations that have been developed to predict ephemeral gully erosion. The available empirical equation by Thorne (1984) is still preliminary and untested. Available theoretical relationships have been proposed by Foster and Lane (1983) and Hirschi and Barfield (1984). The theoretical equations are based on two propositions:

(1) Detachment is proportional to shear excess.

(2) Critical tractive force and channel properties are constant along a channel.

The model of Hirschi and Barfield (1984) includes a simple algorithm for channel wall sloughing. Neither model includes a procedure for headwall or knickpoint advances, nor do they consider stochastic variability. The theoretical relationships have been given only limited validation. Input parameters for the models are virtually nonexistent (Hirschi, 1985). Based on the field observation of these researchers, the models need to be modified to accommodate the following realities:

- Channel properties are not uniform along a given reach. In fact, these nonuniformities may lead to the formation of the head-cut or knickpoint.
- During the formation of ephemeral gullies, the nonuniformity of the channel properties results in a series of chutes and pools. Detachment in this case tends to be more a scour process than resulting from classic shear excess. Thus, the shear excess model may be inappropriate under these conditions.
- During rainfall events, channel growth prior to reaching an impervious layer is influenced by channel wall sloughing, thus an adequate model of channel erosion must account for sloughing. Antecedent moisture conditions must also be taken into account.

Since the available models do not adequately account for these factors and since adequate information for input variables is not available, considerable research is needed before a well-tested operational algorithm is available.

REFERENCES

Foster, G. A. 1986. Understanding ephemeral gully erosion. (This publication).

Foster, G. A., and L. J. Lane. 1983. Erosion by concentrated flow in farm fields. Pp. 9.65-9.82 in Proc. of the D. B. Simons Symposium in Erosion and Sedimentation, Colorado State University, Ft. Collins, Col.

Hirschi, M. C. 1985. Modeling soil erosion with emphasis on steep slopes and the rilling process. Ph.D. dissertation. Department of Agricultural Engineering, University of Kentucky, Lexington, Ky.

Hirschi, M. C., and B. J. Barfield. 1984. Modeling channel erosion with emphasis on upland areas. Paper No. 84-2549. St. Joseph, Mich.: American Society of Agricultural Engineers.

Thorne, C. R. 1984. Prediction of soil loss due to gullies in arable fields. Report CER83-84. Ft. Collins, Colo.: Colorado State University.

The investigation reported in this paper (#85-2-217) is in connection with a project of the Kentucky Agricultural Experiment Station and is published with the approval of the director of the station.

5

Wind Erosion

Dale A. Gillette

Although there was much interest in wind erosion of agricultural soils in the 1930s, research has not been extensive except in a few places including the Great Plains and Southern High Plains. In an effort to include wind erosion data with other kinds of soil erosion data for the entire United States, the Soil Conservation Service (SCS) estimated total wind erosion by using National Resources Inventory (NRI) data and a wind erosion equation that was described by Skidmore and Woodruff (1968). The equation for expressing expected wind erosion is:

$$E = IKCf'(L')V, \tag{1}$$

where I is expected erosion (tons/hectare/year) for a flat bare soil, K is the ridge roughness factor, C is a climatic factor, f'(L') is a function of the fetch length L', and V is the vegetative factor.

The Wind Erosion Equation (WEE) was developed from wind tunnel work, laboratory experimentation, and field observations in the Great Plains and Southern High Plains. For the SCS application, the factor I was estimated from soil texture data that complemented other NRI data; C was obtained from maps; and f'(L'), K, and V were calculated using the field data of the NRI.

Estimations based on Equation 1 and the NRI data proved, however, to be somewhat puzzling. For example, wind erosion for south central Minnesota and northwest Florida (see Table 1) show that wind erosion is of the same magnitude as rill and sheet erosion by water. Intuitively, given soil characteristics, climate, and vegetation in these regions, the results suggest an overestimate of the magnitude of wind erosion. For this

TABLE 1 Wind and Sheet and Rill Erosion for Four Major Land Resource Areas (MLRAs)

MLRA	Location	Wind Erosion (10^6 Ton)	Sheet/Rill Erosion (10^6 Ton)
72	Western Kansas	55	21
77	Western Texas	330	27
103	South central Minnesota	67	57
154	Northwest Florida	0.7	0.9

SOURCE: W. E. Larson, University of Minnesota, personal communication (1984).

reason, an analysis was initiated of the wind erosion equation and the data used in it to see whether features of the equation or of the data would lead to overestimations of wind erosion for localities having soil characteristics and climate different from the area in which the equation was developed [western Kansas, Major Land Resource Area (MLRA) 72].

This paper includes an analysis of four of the five terms used in the WEE, a possible alternative WEE that would correct some perceived shortcomings of the original equation, a proposed provisional WEE (requiring more research for implementation) that may be used with data from the 1982 NRI, and a review of some work on the portion of the eroded soil (dust) that is carried far from the eroded field, to emphasize its importance and to urge that work be started to estimate the loss of this fine portion of the soil. It has potential impact not only on soil but also on atmospheric pollution.

ANALYSIS OF TERMS OF THE WIND EROSION EQUATION

Erodibility Term, I

The cornerstone of the WEE in explaining variance is the soil erodibility term, I. It was derived from the total annual erosion (mass/area/year) for certain farm fields located near Garden City, Kansas.

In the WEE, the I values are used as expected annual wind erosion for soils having a certain parameter. The soil parameter to which this total expected erodibility was related is the percentage of soil mass in aggregates smaller than 0.84 mm. Previous work on threshold velocities that included a regression with percentage of mass smaller than 1 mm (a size very close to 0.84 mm) allowed the expected annual soil loss per unit area (I') to be calculated and compared with the equivalent I factors of the original WEE.

The expected total annual soil loss per unit area could be expressed as an expectation integral divided by field length. The proposed integral is composed of a function that expresses horizontal soil mass flux at the downwind edge of a field as a function of wind friction velocity, a probability density function for wind friction velocity at a given location, and a threshold friction velocity at which erosion starts.

$$I' = \Delta T \int_{u_{*t}}^{\infty} q(u_*)\, f(u_*)\, du_*/L, \qquad (2)$$

where ΔT is the sampling time (1 year for the WEE), L is field length, u_* is a wind friction velocity, u_{*t} is threshold friction velocity, $q(u_*)$ expresses total soil horizontal mass flux (i.e., soil movement) as a function of wind friction speed (u_*), and $f(u_*)$ is the probability density of the wind friction velocity.

H. Lettau (University of Wisconsin, personal communication, 1973) gives the horizontal mass flux [mass/(width • time)] as:

$$q(u_*) = k u_*^2 (u_* - u_{*t}), \qquad (3)$$

where k is a constant. The fit of field data (Gillette, 1981) to this function is shown in Figure 1. Soils 1, 2, 3, 4, and 5 (sand and loamy sand) all have q versus u_* data that fit Equation 3 quite well for threshold friction velocities u_{*t} between 20 and 40 cm/s. These values of u_{*t} for sand and loamy sand textures are quite consistent with outdoor wind tunnel tests for threshold friction velocities. The data points for soil 6 (a sandy loam soil) and soil 9 (a clay textured soil)

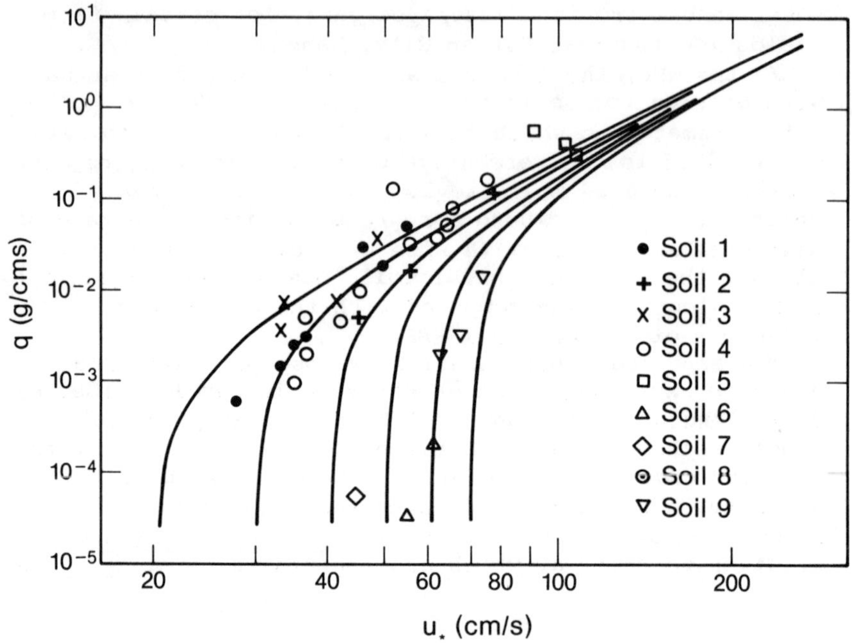

FIGURE 1 Plot of the function $q(u_*) = 4 \times 10^{-7} u_*^2 (u_* - u_{*t})$ versus friction velocity, u_*, and field data (Gillette, 1981). (Textures of soils 1-9 are plotted in Figure 8.) Threshold velocities (u_{*t}) for the six curves (left to right) are 20, 30, 40, 50, 60, and 70 cm/s.

fall close to the curve for u_{*t} = 62 cm/s, which is also consistent with outdoor wind tunnel data for threshold friction velocities. The value for soil 7 (a loamy sand soil) falls on a curve for u_{*t} = 45 cm/s.

A probability density of wind speed, the Rayleigh distribution, which has been used by researchers in the wind energy field (see, for example, Corotis et al., 1978), was used in the expectation integral. Details of the Rayleigh distribution are given in Appendix A. For the parameter of the Rayleigh distribution, data for Dodge City, Kansas (which is located near Garden City, Kansas) were used. Substitution into Equation 2 of the mass flux function of wind speed (Equation 3), the distribution of wind speeds, and the lower limit of wind speed at which the soil erodes gives an expression of expected wind erosion for a flat bare soil:

$$I' = k c_d^{1.5} \, \Delta T \int_{U_t}^{\infty} U^2 (U - U_t) \, f(U) \, dU/L, \tag{4}$$

where c_d equals $(u_*/U)^2$ (the drag coefficient), U is wind speed at 7 m, and f(U) is a Rayleigh probability density function. Therefore, as derived in Appendix A,

$$I' = k c_d^{1.5} \, \Delta T \, [\bar{U}^3 F(x) - \bar{U}^2 U_t G(x)]/L, \tag{5}$$

where x equals $U_t/\bar{U}$, $\bar{U}$ is the mean annual wind speed, U_t is threshold wind speed, and F and G are functions of x (evaluated in Appendix A).

For this calculation, threshold velocities given by Gillette et al. (1980), shown in Figure 2 and corrected to the height of 7 m, were used.

Actually, percentage of mass smaller than 1 mm is not the best predictor of threshold velocity, although it does allow us to compare the annual expected wind erosion with the I factor of the original WEE, which uses the common parameter, percentage of soil mass smaller than 0.84 mm. This comparison is shown in Figure 3. Both curves have been normalized by the expected erosion for all soil mass smaller than 0.84 mm. Thus, both curves represent a relative erosion as a function of percentage of soil mass smaller than 0.84 mm. The agreement is relatively good for high percentage values of soil mass smaller than 0.84 mm. But in the region of less than 70 percent, there is significant disagreement. However, as Chepil (1960) pointed out, this part of the I curve is doubtful: "In view of great inaccuracies in measuring relatively small annual soil losses from depth of soil removal, conversion of the relative field erodibility to annual soil loss based on the curve of [his] Figure 1 must be regarded only as highly approximate."

Use of an expectation integral appears justified because of the progress that has been made in the determination of wind speed probability distributions, of the horizontal flux of soil as a function of wind stress, and of threshold friction velocities for the onset of wind erosion.

On the other hand, an examination of Chepil's data on which I versus percentage of soil mass less than 0.84 mm is based shows considerable scatter of the rather sparse data points from which this most important term of the

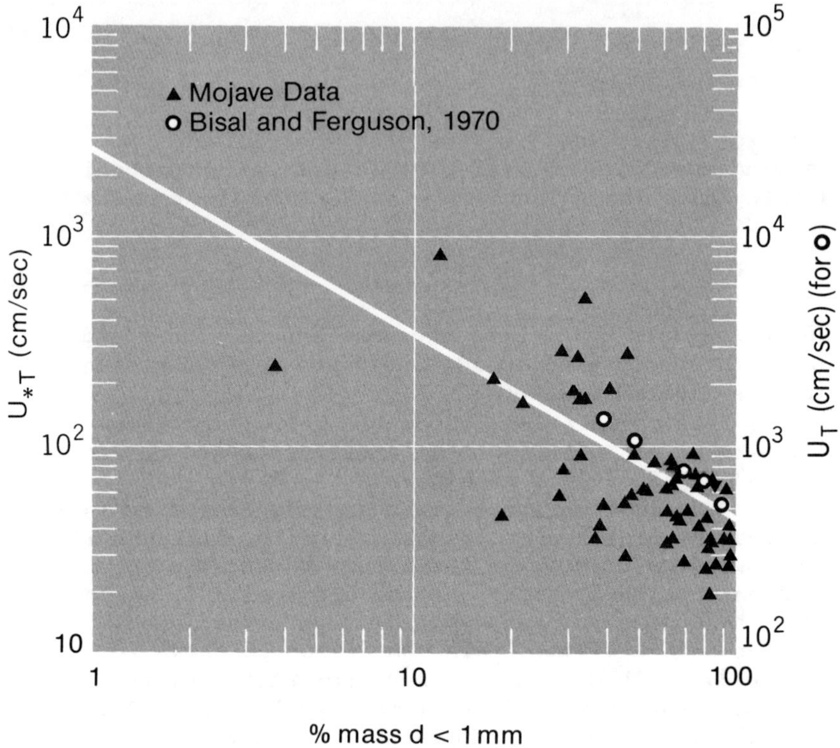

FIGURE 2 Threshold velocities versus percentage of soil mass smaller than 1 mm (Gillette et al., 1980).

WEE was estimated. Data are also relatively sparse for the highest erodibility class. Therefore, Chepil's warning on the reliability of the I function for low erodibility cases should probably be heeded. Indeed, the significant differences of erodibility for the WEE and for the expectation integral based on threshold velocity would indicate an overestimation of erosion using the WEE if the integral method is more correct.

Estimate of I Based on Wind Erosion Groups

The NRI estimates of wind erosion used tables showing values of I for the various subsets of the soil texture domain wind erosion groups (WEG) rather than using dry sieving. (See Table 2 for a typical table of I versus

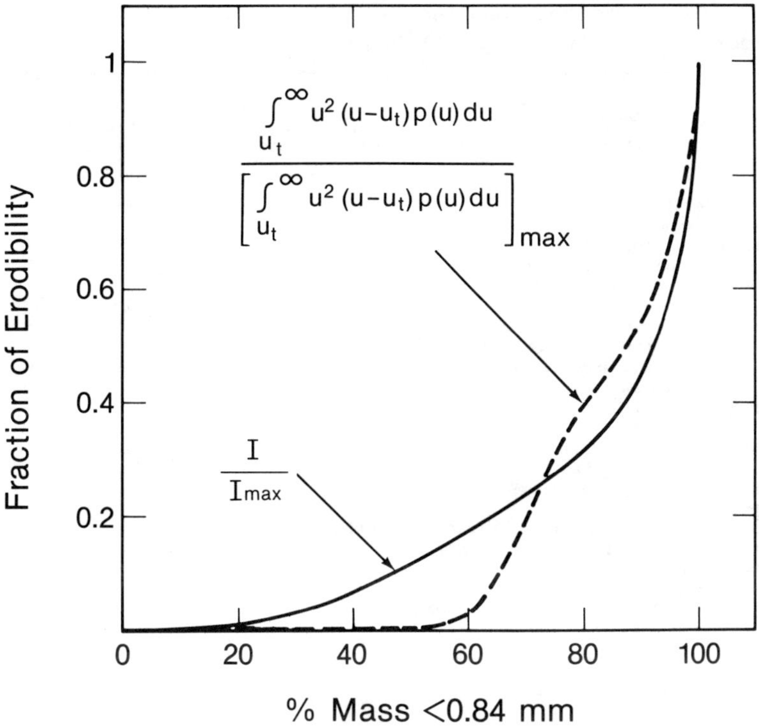

FIGURE 3 Plot of the relative shapes for I curve used in the NRI calculations and the expectation integral of Equation 5 in this paper versus percentage of soil mass in aggregates smaller than 0.84 mm.

WEG and for definitions of WEGs.) These "typical" data were obtained from Lyles (1976) but are not necessarily the same as those used for specific MLRA units to obtain wind erosion estimates (T. George, Soil Conservation Service, personal communication, 1984).

When some of the threshold velocity data for sand- and clay-textured soils (Gillette et al., 1980) were examined, the variability of u_{*t} from which expected erodibility may be calculated was striking. For disturbed sand-textured soils, the mean and standard deviation of threshold friction velocity for seven soils was 31.6 $\pm$ 8.2 cm/s. For disturbed gravelly soils having sand textures, threshold friction velocity was 61 $\pm$ 20 cm/s.

Four clay-textured soils had a mean threshold friction velocity of 31.3 $\pm$ 8.2 cm/s, i.e., practically the same

TABLE 2 Descriptions of Wind Erodibility Groups (WEG) and Corresponding Erodibility (I) Values

WEG	Predominant Soil Textural Class	Dry Soil Aggregates 0.84 mm (Percentage)	Soil Erodibility I [(T/Ha)/Yr]
1	Very fine, fine, and medium sands: dune sands	1	696
2	Loamy sands; loamy fine sands	10	301
3	Very fine sandy loams; fine sandy loams; sandy loams	25	193
4	Clays; silty clays; noncalcareous clay loams; silty clay loams with more than 35% clay content	25	193
4L	Calcareous loams and silt loams; calcareous clay loams; silty clay loams with less than 35% clay content	25	193
5	Noncalcareous loams and silty loams with less than 20% clay content; sandy clay loams; sandy clay	40	126
6	Noncalcareous loams and silt loams with more than 20% clay content; noncalcareous clay loams with less than 35% clay content	45	108
7	Silts; noncalcareous silty clay loams with less than 35% clay content	50	85

SOURCE: Lyles (1976).

value as for the disturbed nongravelly, sand-textured soils. However, three of these four soils had very limited reservoirs of erodible material. Once a small amount of soil had eroded, their threshold velocities returned to high values, which rendered the soil virtually unerodible. The fourth clay soil, a vertisol, had a deep reservoir of erodible material and was capable of eroding as much as the sand-textured soils. Four other disturbed clay-textured soils had friction velocities well above 100 cm/s and were considered almost unerodible.

The great variability of I values will not be explained solely by a relationship with soil texture. This unexplained variability will probably be greater for soil textures other than sand and loamy sand, although the latter may have considerable variability if gravel is a significant constituent of the surface soil layer.

Constancy of I

The use of a constant value for I for an entire year seems inadvisable when considering the changes observed in dry aggregate size distribution for certain textures of soil during one 9-month period. For example, the histograms in Figure 4 show dry aggregate size distributions of a clay-textured soil at the beginning of a drought season in West Texas and at the end of that season 9 months later. The aggregates show a disintegration that led to an increase in I by a factor of about 5. The value of threshold wind speed also changed during this time from 204 cm/s to 35 cm/s.

C Term--Erodibility Corrected for Climate

Effect of Mean Wind Speed Difference

In correcting for areas having different wind speed and rainfall climates (using Garden City, Kansas, as the reference area), the WEE uses a correction based on mean quantities of wind speed and rainfall evaporation. In an attempt to elucidate the effect of this treatment, the climate factor C was simplified by assuming no soil moisture effects; only the effect of differing mean wind speeds in the same way as is done in the WEE was considered.

The C value is used to correct the I value so that erosion can be estimated for an identical farm field located in a different climatic region. Thus, IC would estimate annual erodibility for a farm field that is flat and barren and that has a length and threshold velocity identical to those fields near Garden City, Kansas, where the I values were determined.

Therefore, the question is asked, does a correction factor,

$$C = \bar{U}^3/\bar{U}_{GC}^{\,3}, \quad (6)$$

where the subscript GC stands for the Garden City value of the subscripted variable, properly correct for a change of the distribution for different geographical regions? Evaluation of Equation 5 for Garden City and another location using Appendix A and correcting the Garden City value with Equation 5 yields:

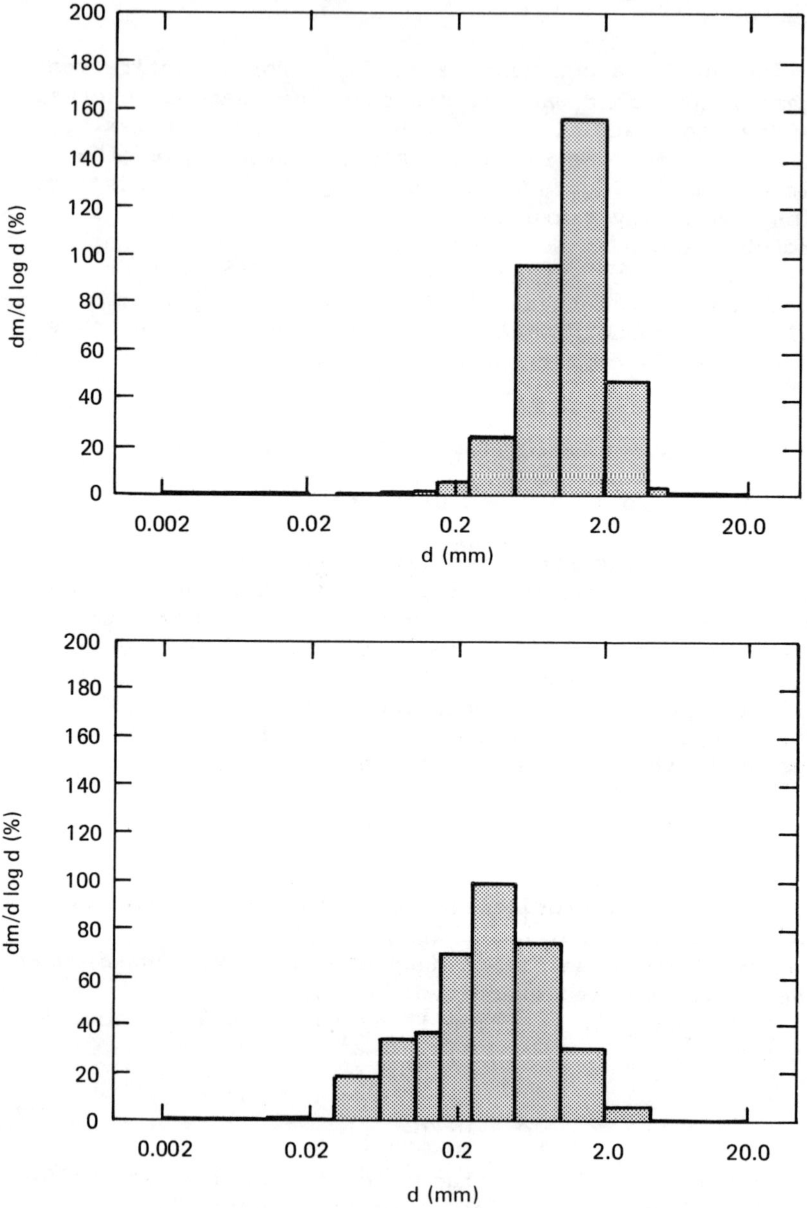

FIGURE 4 Size distributions of loose particles on the surface of a Randall Clay soil (upper) before a season of drought and (lower) after a season of drought.

$$F(x_{GC}) - U_t/\bar{U}_{GC}\ G(x_{GC}) = F(x) - U_t/\bar{U}\ G(x), \qquad (7)$$

where x equals $U_t/\bar{U}$ and x_{GC} equals $U/\bar{U}_{GC}$.

In fact, the evaluation shows that the correction factor C will overestimate wind erosion when the ratio of wind velocity at a particular site to wind velocity at Garden City ($\bar{U}/\bar{U}_{GC}$) is less than one. Thus, the WEE would be expected to overestimate wind erosion for most parts of the United States because the ratio $\bar{U}/\bar{U}_{GC}$ is less than one for most parts of the United States.

This analysis shows that C does not accurately correct the estimate for expected erosion in a different climatic region because mean values of wind speed to the third power do not equal the expectation of the third power of the wind speed. Indeed, a significant overestimation of wind erosion would be expected using C for mean wind speeds that were lower than those at Garden City, Kansas.

Effect of Soil Moisture

According to Chepil (1956), the threshold velocity of soil following moistening is increased by an amount proportional to soil moisture content divided by soil moisture content at 15 bars tension. When the soil dries it will either return to its former physical state and recover its old (lower) threshold velocity or it will form a crust that will determine a new threshold velocity.

Gillette et al. (1982) showed that soil crusts thicker than 1 cm and having a modulus of rupture greater than 1 bar prevent erosion for friction velocities smaller than about 150 cm/s (rarely exceeded by the atmosphere). Thus, until disturbance of the soil disintegrates the surface crust, the soil is for practical purposes unerodible by wind. Crust formation is not prevalent on sandy and sandy loam soils, but it is quite an important mechanism in preventing wind erosion on finer textured soils.

Moreover, the author has observed that wind erosion recurred within minutes of a rainfall for sand and loamy sand soil textures. These observations suggest that crusting of finer textured soils may be more important than soil moisture to wind erosion prevention.

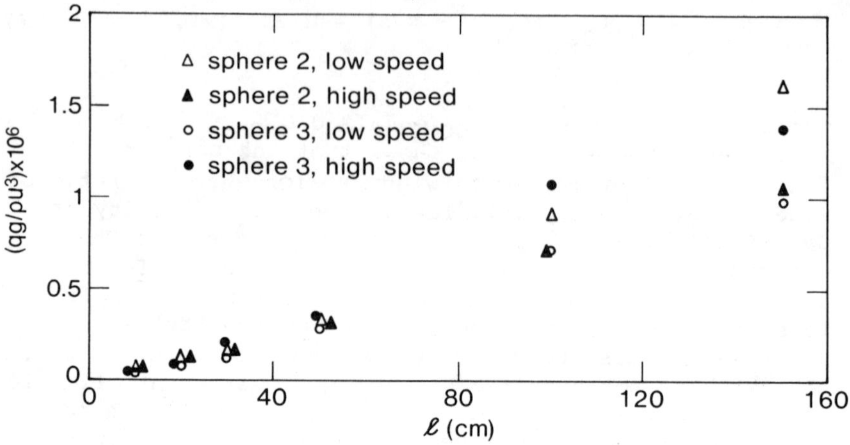

FIGURE 5 Nondimensionalized mass flux, using center-line wind speed as the speed parameter, versus length in a wind tunnel (Gillette and Stockton, 1985).

L' Term--Fetch Length

The wind fetch effect, named "soil avalanching" by Chepil (1957), is an increase with downwind distance of the horizontal flux of soil mass in wind erosion. Actually, avalanching is a misnomer because the increase of soil horizontal flux is not related to conversion of the potential energy of erodible soil particles into kinetic energy. For a constant wind stress and homogeneous soil aggregate structure across a farm field, no increase of soil mass flux with distance should be expected because soil mass flux responds within 10 cm to a change in wind stress (Gillette and Stockton, 1985). An observed increase of particle saltation flux with distance in a wind tunnel has been ascribed to the effect of a feedback mechanism that increases wind stress with distance by increasing the effective aerodynamic roughness height (Owen and Gillette, 1985).

Figure 5 shows the increase of particle flux (expressed as a nondimensional ratio) with distance downwind in a wind tunnel. This increase was accompanied by an increase of the ratio of friction velocity to center-line wind velocity with distance (Gillette and Stockton, 1985) such that the ratio of particle flux divided by friction velocity to the third power remained more or less constant.

The experiment from which Figure 5 was obtained showed that the particular fetch effect observed can be explained by a purely aerodynamic effect. Although it is true that length scales differ greatly between wind tunnel and farm field, this evidence suggests that Chepil's assumption that the wind stress is constant for a given eroding field and his representation of the increase of saltation flux as a function only of soil erodibility is incorrect.

A correct treatment of the fetch effect must consider the roughness of the surface upwind of the eroding field, the height of the planetary boundary layer, the roughness of the eroding field, and the dry aggregate structure of the eroding soil, among other parameters. More work needs to be done on the problem of the fetch effect, especially in the light of recent findings on wind erosion in wind tunnels.

K Term--Ridge Roughness Factor

Ridges or furrows in farm fields affect the flux of eroding particles by establishing an aerodynamic roughness height and by trapping sand in the furrows. Sand trapping appears to be the dominating effect and a deep furrow would be very effective in limiting sand flow on a farm field. For this reason the K factor used in the WEE that expresses fraction of eroding material for a furrowed field to that in a flat field is puzzling. That is, the fraction, K, after a minimum for ridge roughness of about 88 mm, increases with increasing furrow depth (see Figure 6). The increase with increasing furrow depth (see Figure 6) after a minimum furrow is also puzzling. This increase would seem to be explainable only by increased roughness height and extremely erodible soil.

The newer data of Fryrear (1984) (see Figure 6) do not show an upturn in the K factor curve when ridge roughness is greater than 88 mm. The differences in the curves are probably explained by the differences in experimental methods. Whereas Armbrust et al. (1964) set the base of their soil ridges even with the bottom of the wind tunnel and used highly erodible dune sand and gravel, Fryrear (1984) set his wind tunnel 20 mm below the peak of the ridges. Fryrear's method may have simulated field conditions better because it simulates the action of a boundary layer that has its point of zero mean wind speed not too far from the tops of the soil ridges. Substitution of Fryrear's K factor in the WEE would lower the

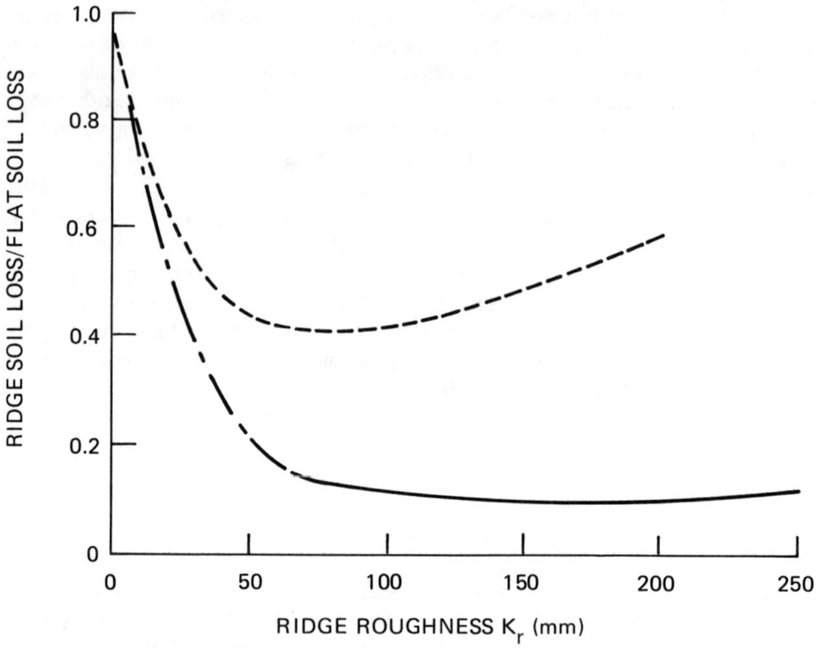

FIGURE 6 Ratio of soil loss from ridged surface to soil loss from a flat surface with the same soil versus ridge roughness. Data from Fryrear (1984) are a solid line; those from Armbrust et al. (1964), a dashed line. Results from Chepil and Doughty (1939) are marked Y; from Woodruff et al. (1968), X; and from Fryrear and Armbrust (1969), O. (Entire figure from Fryrear, 1984.)

total estimated erosion by lowering the estimates for deep-furrowed farm fields.

The possibility that field furrow depth can change during the erosion season makes the approach of using only one value of K for the entire season a probable source of error.

V Term--Vegetative Effect

Nonerodible material on the surface acts to limit erosion in two ways. The first and most obvious way is that the soil surface is covered and thus not exposed to erosive forces. Second, nonerodible elements partition the wind stress in such a way that a fraction of the stress is absorbed by those elements, leaving a residual wind stress to erode the erodible soil lying in between. The Wind Erosion Unit of the Agricultural Research Service has been using this conceptual framework of momentum partitioning in assessing the effect of vegetation and has made good progress, as described, for example, by Lyles and Allison (1976).

The incorporation of this kind of work into the new WEE proposed in the next section would be beneficial. In this equation, the threshold velocity would be determined not only by the physical state of the exposed soil, but also by partitioning of momentum by nonerodible elements (for example, vegetation) and other effects (for example, soil moisture and trapping of particles by furrows and surface residues).

The wind momentum flux available for wind erosion of soil would continue to be affected by nonerodible elements after threshold velocity is increased and that effect would be included in the drag coefficient c_d. Again, any change of the vegetative cover should be represented in the new wind erosion equation by a change of the threshold velocity and drag coefficient.

Knoll Erodibility

Increased flux of soil particles on an upslope may be explained by considering the physics of saltation. Saltation is a type of particle motion by which particles move through the air by jumps and return to the surface. An upslope probably causes the saltating particle to have a shorter flight length. A more detailed theoretical analysis of particle movement is needed to produce a better estimate of knoll erodibility.

AN ALTERNATIVE WIND EROSION EQUATION

As this discussion indicates, the present WEE leads to possible overestimation of annual wind erosion loss by its structure and dependence on a limited data set. Recent work since the formulation of the equation by Skidmore and Woodruff (1968) suggests that many of the factors in the original equation need to be reevaluated. An approach using the expectation integral given in Equation 2 appears to be superior to the original wind erosion equation, Equation 1, because it more closely follows the aerodynamics and physics of wind erosion. That is, it has as its basis verified distributions of wind speed, saltation flux as a function of wind stress, and wind threshold velocity, and it combines them in a way consistent with experimentation in the physics of wind erosion.

A proposed alternative WEE is the sum of n expectations for n periods of time, which added together cover the period of interest.

$$E' = k \sum_{i=1}^{n} (c_{d_i}^{1.5}) r_i \, \Delta T_i \int_{U_{t_i}}^{\infty} U^2 (U - U_{t_i}) p_i(U) \, dU/L, \qquad (8)$$

where each period of time ΔT_i (in seconds) represents a period when the parameters remain relatively constant; $p_i(U)$ is the probability density of wind speed during i; $U_{t_i} = u_{*t_i} c_{d_i}^{-0.5}$, which is affected by soil aggregation, vegetation, soil moisture, and ridge roughness; r_i is the fraction of ground not covered by vegetation or other nonerodible elements; c_{d_i} is the drag coefficient for time period i ($c_d = [u_*/U]^2$); L is field length (meters); k is a constant; and U is a wind speed at 7 m.

Equation 8 is a version of the expectation integral given in Equation 2 and has as its variables the threshold wind speed, drag coefficient, fraction of surface covered, field length, and a constant coefficient. All the variables considered by the original WEE are implicit in the variables of this alternative. Standing vegetation and vegetative residue would affect the fraction of surface covered and partitioning of wind stress by the nonerodible elements (the vegetation and vegetative residue), and the erodible soil would affect the threshold friction velocity. Threshold wind speed would also be affected by soil aggregate structure, crusting, and soil moisture. The field length effect, ridge roughness, and vegetation-vegetative residue would affect the drag coefficient and sand trapping. The integral would be evaluated for every significant change of threshold velocity, drag coefficient, wind probability distribution, and fraction of surface covered.

Such a formulation would be responsive to changes of soil conditions and would treat erosion in a manner consistent with experimental and theoretical work in wind erosion. As experience was gained with the new wind erosion equation, changes could be easily incorporated and simplifications easily made. For example, improvements for such effects as field length, changing of the threshold velocity as a function of soil conditions, and possibly a revised knoll effect would probably improve the wind erosion estimates.

A PROPOSED "PROVISIONAL CORRECTED WEE" FOR COMPUTATION OF WIND EROSION FROM THE 1982 NRI

In light of the above commentary on the original wind erosion equation and the proposed form for a new wind erosion equation (Equation 8), which will require much research to implement, an incomplete "provisional wind erosion equation" is suggested here for estimating wind erosion from the 1982 NRI data. The proposed provisional equation corrects some of the shortcomings of the original WEE but will not correct for many effects that will require much more research. Corrections include replacement of the I and C factors of the original wind erosion equation with the expectation formula given in Equation 4 and replacement of the old ridge roughness factor K with Fryrear's (1984) ridge roughness factor.

The provisional corrected wind erosion equation does not use the series of values for c_d, U_t, r, p, and ΔT as given in Equation 8 because these values are not in the 1982 NRI data, and research is lacking to complete formulation of the new wind erosion equation. Rather, one value for each parameter is used for the entire year, and the old field length formulation and old vegetation factors will be used. It will be noted that the soil moisture effect is also temporarily ignored. The soil flux function of Equation 3 is used along with the Rayleigh probability density of Appendix A.

A value for c_d of 0.002 is assigned based on a selection of field measurements on eroding soils. This value is consistent with drag coefficients given by Priestly (1959) for similar surfaces. The combined proposed provisional wind erosion equation for 1 year is given below:

$$E' = 1{,}127\ [\bar{U}^3 F(x) - \bar{U}^2 U_t G(x)]\ K f'(L') V/L\ [(t/ha)/yr], \quad (9)$$

where x equals $U_t/\bar{U}$, V and f(L) are as in the original WEE (Equation 1), K is the ridge roughness factor of

Fryrear (1984), F(x) and G(x) are given in Appendix A, and the remaining variables are as designated for Equation 5.

Since the last three factors of this new provisional WEE are either equal to or evaluated similarly to those in the original WEE, and field length is already measured in the 1982 NRI, only the variables $\bar{U}$ and U_t and their ratio $U_t/\bar{U}$ must be determined to use this provisional WEE. Values for $\bar{U}$ are given in Table 3 for selected locations in the United States. Values for U_t are given in Table 4 for seven WEGs as defined in Table 2. These values were simply based on the percentage of soil mass smaller than 0.84 mm as given for the seven WEGs by Lyles (1976) in Table 2, and the values of U_{*t} given in Figure 2 corrected for height to give U_t. Functions F(x) and G(x) are given in Appendix A.

Table 5 gives some sample results for the proposed provisional WEE before the Kf(L)V correction factors are applied. Values for erodibility,

$$I' = \{1{,}127\ [\bar{U}^3 F(x) - \bar{U}_t^2 U\ G(x)]\}/L \qquad (10)$$

for 1,127-m-long flat bare fields (somewhat longer than those used by Chepil in his 1960 work) are given in Table 5 along with WEGs for Dodge City, Kansas (near the location of Chepil's data source), and for Minneapolis, Minnesota (a location where it was felt that the old wind erosion equation could be overestimating wind erosion).

A comparison of the erodibility values (I') for Dodge City in the present work with the I values versus WEG given by Lyles (1976) shows moderately good agreement for the low numbered WEGs but large disagreement for WEGs 6 and 7. However, comparison of erodibility values (I') for Minneapolis with Dodge City values (R • I') corrected by $(\bar{U}_{Minneapolis}/\bar{U}_{Dodge\ City})^3$ (i.e., the kind of correction used by the original wind erosion equation) shows that the R • I' is larger by a factor of 2 for WEGs 1 and 2; by a factor of 5 for WEGs 3, 4, and 4L; and by about a factor of 10 for WEGs 5 and 6.

The substitution of Fryrear's (1984) ridge roughness factor for the K factor of the original WEE will give lower estimates for fields having deep furrows. It is not known how important the neglect of soil moisture will be, although it can reasonably be said that the estimates will be a bit high.

TABLE 3 Mean Wind Speed ($\bar{U}$) for Selected U.S. Stations

Station	State	$\bar{U}$ (M/S)	Station	State	$\bar{U}$ (M/S)
Birmingham	AL	3.3	Detroit	MI	4.6
Montgomery	AL	3.0	Grand Rapids	MI	4.5
Tucson	AZ	3.7	Lansing	MI	4.6
Yuma	AZ	3.5	Sault St. Marie	MI	4.3
Fort Smith	AR	3.4	Duluth	MN	5.1
Little Rock	AR	3.6	Minneapolis	MN	4.7
Fresno	CA	2.8	Jackson	MS	3.4
Red Bluff	CA	3.9	Columbia	MO	4.4
Sacramento	CA	3.7	Kansas City	MO	4.6
San Diego	CA	3.0	St. Louis	MO	4.2
Denver	CO	4.1	Springfield	MO	5.0
Grand Junction	CO	3.6	Billings	MT	5.1
Pueblo	CO	3.9	Great Falls	MT	5.9
Hartford	CT	4.0	Havre	MT	4.5
Washington	DC	3.4	Helena	MT	3.5
Jacksonville	FL	3.8	Missoula	MT	2.7
Tampa	FL	3.9	North Platte	NE	4.6
Atlanta	GA	4.1	Omaha	NE	4.8
Macon	GA	3.5	Valentine	NE	4.8
Savannah	GA	3.6	Ely	NV	4.7
Boise	ID	4.0	Las Vegas	NV	4.0
Pocatello	ID	4.6	Reno	NV	2.9
Chicago	IL	4.6	Winnemucca	NV	3.5
Moline	IL	4.4	Concord	NH	3.0
Peoria	IL	4.6	Albuquerque	NM	4.0
Springfield	IL	5.1	Roswell	NM	4.1
Evansville	IN	3.7	Albany	NY	4.0
Fort Wayne	IN	4.6	Binghamton	NY	4.6
Indianapolis	IN	4.3	Buffalo	NY	5.5
Burlington	IA	4.6	New York	NY	5.5
Des Moines	IA	5.0	Rochester	NY	4.3
Sioux City	IA	4.9	Syracuse	NY	4.4
Corcordia	KS	5.4	Cape Hatteras	NC	5.1
Dodge City	KS	6.3	Charlotte	NC	3.4
Topeka	KS	4.6	Greensboro	NC	3.4
Wichita	KS	5.6	Wilmington	NC	4.0
Louisville	KY	3.8	Bismarck	ND	4.7
Shreveport	LA	3.9	Fargo	ND	5.7
Portland	ME	3.9	Cleveland	OH	4.8
Baltimore	MD	4.2	Columbus	OH	3.9
Boston	MA	5.6	Dayton	OH	4.6

TABLE 3 (Continued)

Station	State	$\bar{U}$ (M/S)	Station	State	$\bar{U}$ (M/S)
Toledo	OH	4.2	Dallas	TX	4.9
Oklahoma City	OK	5.7	El Paso	TX	4.2
Tulsa	OK	4.7	Port Arthur	TX	4.5
Portland	OR	3.5	San Antonio	TX	4.2
Harrisburg	PA	3.4	Salt Lake City	UT	3.9
Philadelphia	PA	4.3	Burlington	VT	3.9
Pittsburgh	PA	4.2	Lynchburg	VA	3.5
Scranton	PA	3.8	Norfolk	VA	4.7
Huron	SD	5.3	Richmond	VA	3.4
Rapid City	SD	5.0	Quillayute	WA	3.0
Chattanooga	TN	2.8	Seattle	WA	4.1
Knoxville	TN	3.3	Spokane	WA	3.9
Memphis	TN	4.1	Green Bay	WI	4.6
Nashville	TN	3.6	Madison	WI	4.4
Abilene	TX	5.4	Milwaukee	WI	5.3
Amarillo	TX	6.1	Cheyenne	WY	5.9
Austin	TX	4.2	Lander	WY	3.1
Brownsville	TX	5.3	Sheridan	WY	3.6
Corpus Christi	TX	5.4	Elkins	WV	2.8

SOURCE: Department of Commerce (1977).

LONG-DISTANCE LOSS OF SOIL MATERIAL

It would be a worthy goal for estimates of soil removal to be made for that portion of the eroded soil that is carried far from the location of erosion. Eroded soil moving in saltation and creep (movement having lower trajectories than saltation) is removed from a particular farm field, but it is often deposited in a nearby location. It may even be restored to the original field should an equally strong wind from the opposite direction happen to erode the deposited material.

Fine soil material that is carried in suspension, however, has the potential of being carried great distances from the eroding field and being lost to an entire agricultural region. Because this fine soil material is associated with field soil moisture capacity and with important nutrients, its loss may be far more

TABLE 4 Selected Values of Threshold Wind Speed U_t Versus Wind Erosion Group (WEG)

WEG	U_t (M/S)
1	6.6
2	7.7
3	11.1
4	11.1
4L	11.1
5	13.3
6	17.7
7	19.9

NOTE: See Table 2 for a description of wind erosion groups.

SOURCE: Based on Lyles (1976).

important than the loss of the coarser particles that are moved in saltation and creep. The fine-grained sediment carried from fields represents a potential offsite impact as an air pollutant, and upon settling onto surfaces it may have damaging offsite effects.

The original WEE refers to total soil loss to a farm field and not to the loss of the fine portion of the soil. This loss of fine soil carried in suspension by the air may approach in magnitude the loss of saltation/creep-transported soil, but it cannot be simply calculated as a constant fraction of the estimated soil loss given by the wind erosion equation. Rather, it must be considered as a function of wind erosion fluxes over the entire field and in individual erosion events. As the proposed WEE is constructed, it would be suitable for use in estimating the long-distance loss of soil material.

Particles that are carried great distances have fall velocities that are a small fraction of the friction velocity (an approximate scale for the root-mean-square vertical velocity fluctuations of the air). Thus, for a given wind speed and drag coefficient, the maximum size of particle that may be lost to an agricultural region can be calculated (<u>see</u> Figure 7).

TABLE 5 Soil Wind Erodibility[a] (I') near Dodge City, Kans., and Minneapolis, Minn., for 1,127-m-Long Flat Bare Fields, Compared with I Values from Lyles (1976)

WEG	I (Lyles, 1976)	I' Dodge City (T/Ha)/Yr New Provisional Equation	R Times I' for Dodge City[b] (T/Ha)/Yr	I' for Minneapolis (T/Ha)/Yr New Provisional Equation
1	696	434	180	111
2	301	354	147	73
3	193	137	57	10
4	193	137	57	10
4L	193	137	57	10
5	126	57	24	2
6	108	5	2	0.01
7	85	1	0.5	0.001

[a]Erodibility $(I') = \frac{1{,}127}{L} [\bar{U}^3 F(x) - \bar{U}^2 U_t G(x)]$ [(t/ha)/yr].

[b]$R = \left[\frac{\bar{U}_{Minneapolis}}{\bar{U}_{Dodge\ City}}\right]^3$ (a simple correction factor for using Dodge City erosion to estimate Minneapolis erosion).

At a given point in an eroding field, the horizontal mass flux q' of saltating particles and creeping particles moving through a surface perpendicular to the ground and to the wind greatly exceeds the vertical mass flux F'_a of particles carried in suspension that may be transported great distances from the farm field (see Figure 8). Considering, however, that the loss of total soil mass for a given width of eroding field is the horizontal mass flux at the downwind boundary of the field, and that the suspended material portion of this flux is approximately equal to the integrated vertical flux of fine material over the entire area of the eroding field, the loss of fine material clearly becomes a much larger fraction of the total loss of soil mass.

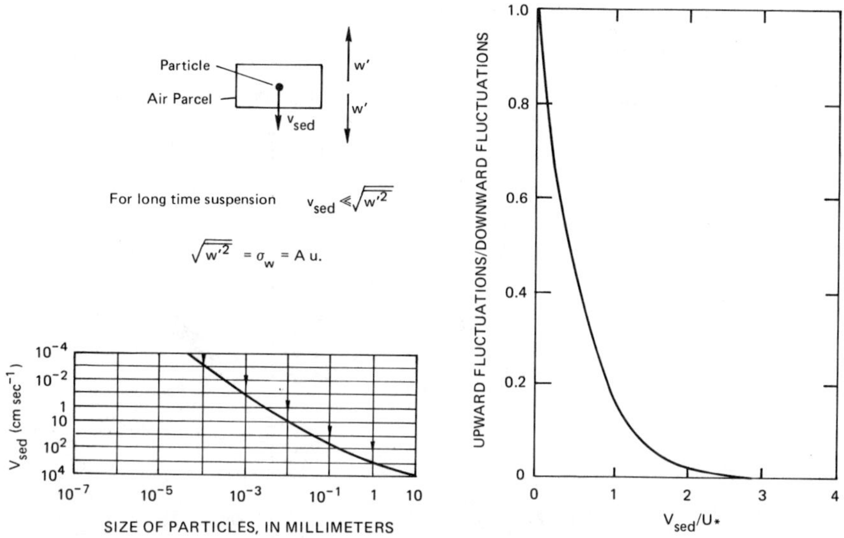

FIGURE 7 Clockwise from upper left: sedimentation velocity V_{sed} compared to vertical velocity fluctuation w'; upward motions divided by downward motions for a particle having a sedimentation velocity V_{sed} in air having vertical velocity fluctuations with mean zero and standard deviation u_*; sedimentation velocity versus particle size. (This part of figure from Bagnold, 1941; entire figure from Gillette, 1981.)

The vertical suspension mass flux is probably strongly related to the horizontal saltation flux. This relationship seems to arise because the kinetic energy flux to the surface by saltating particles is related to the horizontal mass flux of saltating particles (Gillette and Stockton, 1985) and because the production of fine particles by sandblasting is related to the flux of kinetic energy to the surface (Hagen, in press). Indeed, some of the measurements shown in Figure 8 show a striking (though admittedly noisy) constancy between horizontal mass flux of saltating and creeping soil and the vertical flux of particles carried in suspension (Figure 8, bottom). This relationship suggests a method for estimating total loss of fine material for long distances; however, development of the method will require fundamental research.

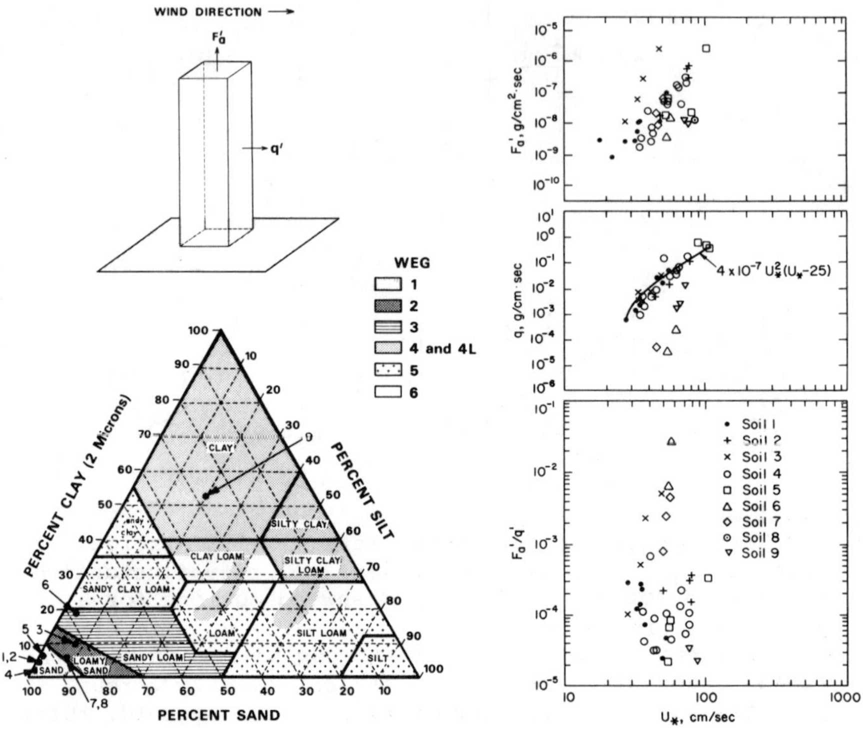

FIGURE 8 Clockwise starting with triangle: textures of sampled soils; illustration of horizontal flux q' and vertical flux F'_a total soil movement versus wind friction velocity; vertical flux of particles smaller than 0.02 mm versus wind friction velocity; ratio of vertical flux of particles smaller than 0.02 mm to total soil movement per unit area per time versus wind friction velocity (from Gillette, 1981).

The vertical flux of fine soil material that is subsequently carried great distances may be estimated by using some of the variables in the NRI data set. However, this estimate will probably be rough and will need much fundamental research to have the same validity as do those for sheet and rill erosion.

CONCLUSIONS

Analysis of the WEE reveals that it probably overestimates wind erosion for values of M smaller than 65 percent where M is percentage of soil mass in aggregates smaller than 0.84 mm. Since small values of M often correspond to the higher numbered wind erosion groups, a systematic error probably exists in erosion estimates for these soils. The method of correcting for mean wind speed also probably leads to an overestimation of wind erosion for most locations in the United States where mean wind speed is less than at Garden City, Kansas. Because soil aggregation can change during the season, the assignment of only one value to potential erosion for an entire season is questionable. New evidence shows that increase of soil mass flux with field length is related to a feedback mechanism that increases aerodynamic roughness height with field length.

An alternative WEE is proposed that would correct for perceived shortcomings of the original WEE. Unfortunately, insufficient data exist to implement this equation, so a provisional WEE is proposed that would improve some of the features of the original equation but still retain some formulation that should be replaced when sufficient research becomes available.

Suspended soil material (dust) also constitutes an important product of erosion. A version of the new WEE may be used to estimate dust emission, but only after several problems relating dust emission and saltation flux are solved.

REFERENCES

Armbrust, D., W. Chepil, and F. Siddoway. 1964. Effects of ridges on erosion of soil by wind. Soil Sci. Soc. Am. Proc. 28:557-560.

Bagnold, R. A. 1941. The Physics of Blown Sand and Desert Dunes. London: Methuen.

Chepil, W. S. 1956. Influence of moisture on erodibility of soil by wind. Soil Sci. Soc. Am. Proc. 19:288-292.

Chepil, W. S. 1957. Width of Field Strips to Control Wind Erosion. Agricultural Experiment Station Technical Bulletin No. 92. Manhattan: Kansas State University.

Chepil, W. S. 1960. Conversion of relative field erodibility to annual soil loss by wind. Soil Sci. Soc. Am. Proc. 24:143-145.

Chepil, W. S., and J. L. Doughty. 1939. Wind tunnel experiments on soil drifting. P. 19 in Report Regional Commission on Soil Drifting, Swift Current, Saskatchewan, July 11.

Corotis, R. B., A. B. Sigl, and J. Klein. 1978. Probability models of wind velocity magnitude and persistence. Solar Energy 20:483-493.

Department of Commerce. 1977. Local Climatological Data--Annual Summaries for 1977. Asheville, N.C.: National Climatic Data Center.

Fryrear, D. W. 1984. Soil ridges--clods and wind erosion. Trans. ASAE 27:445-448.

Fryrear, D. W., and D. V. Armbrust. 1969. Cotton gin trash for wind erosion control. Texas Agricultural Experiment Station, MP-928.

Gillette, D. A. 1981. Production of dust that may be carried great distances. Pp. 11-26 in Desert Dust: Origin, Characteristics, and Effect on Man, T. Pewe, ed. Special Paper 186. Boulder, Colo. Geological Society of America.

Gillette, D. A., and P. H. Stockton. 1985. Mass, momentum and kinetic energy fluxes of saltating particles over eroding particle deposits of varying lengths in a wind tunnel. Unpublished.

Gillette, D. A., J. Adams, A. Endo, and D. Smith. 1980. Threshold velocities for input of soil particles into the air by desert soils. J. Geophys. Res. 85(C10):5621-5630.

Gillette, D. A., J. Adams, D. Muhs, and R. Kihl. 1982. Threshold friction velocities and rupture moduli for crusted desert soils for the input of soil particles into the air. J. Geophys. Res. 87:9003-9015.

Hagen, L. In press. Soil aggregate abrasion by impacting sand and soil particles. Trans. ASAE 27:805-808.

Lyles, L. 1976. Wind erosion: Processes and effect on soil productivity. Paper presented at Annual Meeting of the American Society of Agricultural Engineers, Lincoln, Nebr., June 1976.

Lyles, L., and B. Allison. 1976. Wind erosion: The protective role of standing stubble. Trans. ASAE 19:61-64.

Owen, P. R., and D. A. Gillette. 1985. The wind tunnel effect on wind erosion. Unpublished.

Priestley, C. H. B. 1959. Turbulent Transfer in the Lower Atmosphere. Chicago: University of Chicago Press.

Skidmore, E. L., and N. P. Woodruff. 1968. Wind Erosion Forces in the United States and Their Use in Predicting Soil Loss. Agriculture Handbook No. 346. Washington, D.C.: U.S. Government Printing Office.

Woodruff, N. P., B. L. Schmidt, E. L. Skidmore, J. D. Dickerson, R. L. Meeker, and L. M. Feusner. 1968. A Study of Wind Erosion in Northwestern Ohio. Wind Erosion Unit, U.S. Department of Agriculture, Manhattan, Kans.

APPENDIX: DERIVATION OF EXPECTATION FORMULA

Equation 4

$$I' = kc_d^{1.5} \Delta T \int_{U_t}^{\infty} U^2(U - U_t) f(U) dU/L$$

$$= kc_d^{1.5} \Delta T [\int_{U_t}^{\infty} U^3 f(U) dU - U_t \int_{U_t}^{\infty} U^2 f(U) dU]/L$$

may be rewritten as

$$I' = kc_d^{1.5} \Delta T [I_1 - I_2]/L. \quad \text{(A-1)}$$

Now, let us evaluate I_1, which is given as

$$I_1 = \int_{U_t}^{\infty} U^3 f(U) dU.$$

By substituting

$$f(U) = (\pi U)/(2\bar{U}^2) \exp[-\pi U^2/4\bar{U}^2]$$

(the Rayleigh density function) and letting

$$t = [\pi^{1/2} U/(2\bar{U})]^2,$$

we get

$$I_1 = (4/\pi)^{1.5} \bar{U}^3 \int_z^{\infty} t^{1.5} e^{-t} dt, \quad \text{(A-2)}$$

where z equals $(\pi/4)\ (U_t/\bar{U})^2$.
This has the solution (Abramowitz and Stegun, 1970)

$$I_1 = (4/\pi)^{1.5} \bar{U}^3\ \Gamma(2.5, z), \quad \text{(A-3)}$$

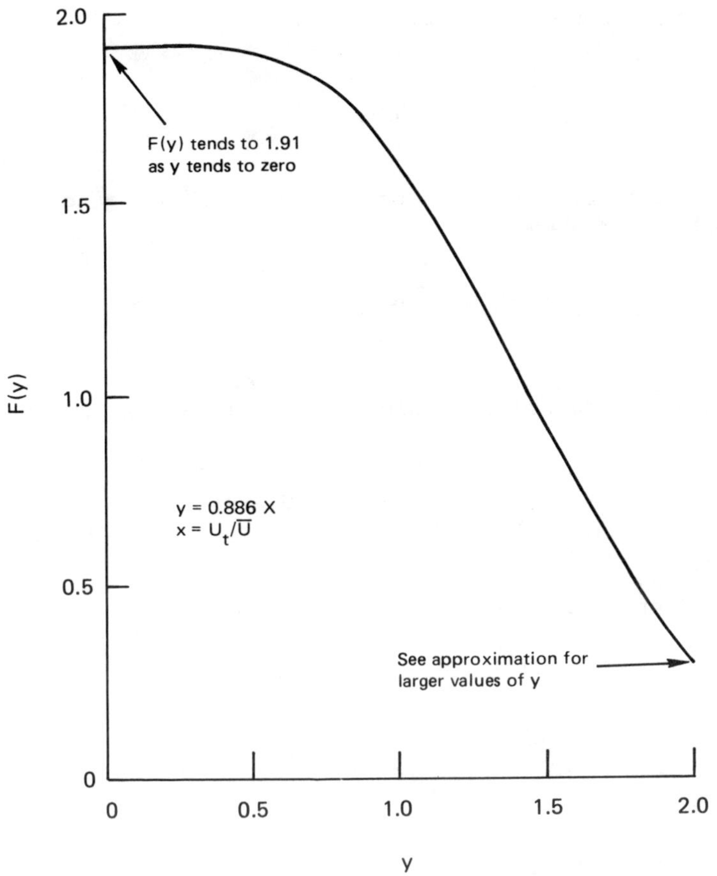

FIGURE A-1 Function F(0.886x) versus 0.886x, where y equals 0.886x (Cowherd et al., 1984).

where $\Gamma(2.5, z)$ is an incomplete gamma function.

The function

$$F(y) = (4/\pi)^{1.5}\ \Gamma(2.5,\ \pi X^2/4) \qquad \text{(A-4)}$$

where x equals $U_t/\bar{U}$ and y equals 0.866x is plotted in Figure A-1 (after Cowherd et al., 1984). For values of x greater than 1.6, the approximation given by Abramowitz and Stegun (1970) is used:

$$F(x) = 1.44[(0.70\ x^3 + 1.33\ x)\ \exp(-\pi x^2/4)]. \qquad \text{(A-5)}$$

Now, the second integral,

$$I_2 = U_t \int_{U_t}^{\infty} U^2 f(U)\, dU,$$

becomes, after substituting for f(U) and changing variables as above,

$$I_2 = 4\, \bar{U}^2\, U_t/\pi \int_{z}^{\infty} t\, e^{-t} dt. \qquad \text{(A-6)}$$

After the integral is evaluated, the equation may be rewritten

$$I_2 = \bar{U}^2\, U_t G(x), \qquad \text{(A-7)}$$

where

$$G(x) = (4/\pi)\, \exp(-\pi x^2/4)\, [(\pi x^2/4) - 1].$$

Thus, equation A-1 may be rewritten

$$I' = k c_d^{1.5}\, \Delta T\, [\bar{U}^3\, F(x) - \bar{U}^2\, U_t\, G(x)]/L \qquad \text{(A-8)}$$

where x equals $U_t/\bar{U}$.

REFERENCES

Abramowitz, M., and I. A. Stegun. 1970. Handbook of Mathematical Functions. New York: Dover Publications.

Cowherd, C., G. Muleski, P. Engelhart, and D. A. Gillette. 1984. Rapid Assessment of Exposure to Particulate Emissions from Surface Contamination Sites. Report to Environmental Protection Agency. Contract No. 68-03-3116, Project No. 7972-L MRI. Kansas City, Mo.

Discussion

Klaus W. Flach

The Wind Erosion Equation (WEE) was developed in the 1950s and 1960s primarily as an operational tool for soil conservationists to evaluate the effects of alternative erosion control practices. For this it has been useful. Despite difficulties cited in Gillette's paper, the formula has been useful in the National Resources Inventory (NRI) to identify those parts of the country--the Great Plains--where erosion by wind is the pervasive soil conservation concern. As pointed out, NRI data suggesting that erosion by wind for parts of Minnesota and Florida exceeds that by water may be questioned. The equation had never been tested in these areas and local Soil Conservation Service (SCS) personnel are relatively inexperienced in its use.

In any case, values for erosion by wind in these areas slightly in excess of erosion by water do not present the picture of an all-important problem related to wind erosion, although adding tons of soil loss by wind and water may result in high assessment of the areas' overall erosion problems.

Although all soil erosion processes deal in principle with the same phenomenon--namely, the movement of soil in response to applied energy--there are major differences between air and water as the source of energy. These differences determine the conditions under which erosion occurs, the accuracy with which it can be measured, and the mechanisms through which erosion influences the quality of soil as a medium for plant growth. Some of these differences are described later herein in a very simplistic way to illustrate that erosion by wind is a much more complex process than that by water, and that equations to predict it will be less precise and results more difficult to interpret until a major research effort is made.

Water is heavier than air and only runs downhill, carrying soil with it. It is relatively simple to measure the amount of soil carried by water. The Universal Soil Loss Equation (USLE) was developed and verified with thousands of such measurements.

Wind, on the other hand, carries eroded soil in one direction on one day and in another direction on another day. The best part of the soil is carried as dust into the air and may move thousands of miles. Some may be blown to the next field and replace some of the soil that has been lost there. There is no easy way to measure the amount of soil moved other than measuring soil thickness before and after the erosion event. This is a very crude measurement that, as Gillette's paper points out, can be used to measure the loss of 1 to 2 inches of soil (150 tons/acres), not the loss of 5 to 10 tons/acre that concerns soil conservationists.

Because air is much less dense than water, a higher fluid velocity is required to move soil by wind. Various particle sizes and aggregates of different densities are moved selectively. Hence, there is a strong winnowing effect of wind, at least at relatively low velocities. As discussed, the deterioration of soil quality because of the preferential loss of fine soil particles may be a more appropriate measure of the effects of erosion by wind than the total amount of soil lost. Very intense wind storms, on the other hand, can remove the entire plow layer of whole fields within hours. Once again, "tons of soil loss" seems an inadequate way to describe such events.

Since soil is always wet when it is being eroded by water, properties that are important for predicting erodibility are relatively constant during the process. On the other hand, a wet medium- or fine-textured soil is virtually immune to erosion by wind until it dries. And when it dries, it commonly forms a crust that protects the soil from further erosion. The WEE tries to account for this in a very simplistic way that, in the days before computers in field offices, was perhaps the only way. For evaluating conservation alternatives for an individual field this correction may have been reasonably adequate, but for a national assessment it is not. [To comment on a point made in this paper, sandy loams and even loamy sands in soils with xeric (Mediterranean climate) moisture regimes can form extremely strong crusts.]

To be truly reliable, the equation must take into account that similar textures may behave quite differently in various soils. At this time no good measure or national data base exists to incorporate the propensity of soils to form crusts into an equation.

Erosion by both water and wind have ephemeral effects on standing crops. Sediments moved by water can damage crops in relatively small areas in the bottom of fields and cause severe damages to streams and lakes. Windblown soil can destroy crops over large areas, form dunes that block roads, and carry pollutants to waterways over large distances. In parts of the United States where erosion by wind is important, many farmers pay as much or more attention to these ephemeral effects as they do to the damage that wind erosion does to the potential productive capacity of the soil. To assess the total damage of erosion, these ephemeral effects must be fully considered.

Wind erosion research has received relatively little attention since 1935, when the problems of the Dust Bowl region awakened the United States to the dangers of soil erosion. Low priorities for research on soil erosion by wind are even reflected in the terminology and agenda for this convocation, "Physical Dimensions of the Erosion Problem," which implies "by water" when using the term erosion. The equation for predicting erosion by water is called the Universal Soil Loss Equation, yet the universe does not include erosion by wind. It is, of course, not wind that erodes, but soil.

The difficulty of conducting research on erosion by wind probably has been a major factor in the low interest of researchers in this subject. Yet NRI data, imperfect as they may be, suggest that erosion by wind may be an extremely important component of the U.S. soil erosion problem.

Gillette's paper provides some excellent suggestions for new directions in research on erosion by wind. The technology is now available to make meaningful measurements of erosion by wind in the field and, through microcomputers and computerized data bases, to provide soil conservationists with the wherewithal to use even a complicated equation in conservation planning.

The difficulties of predicting with current technology the amount and impact of erosion by wind probably preclude the suggestion in this paper to modify the WEE as used with the 1982 NRI. There are essentially no experimental data available for those parts of the country, such as the northern Corn Belt or Florida, where erosion by wind may be significant but is not the dominant form of erosion. Yet these are the areas where the use of the WEE in the NRI is most suspect. Rather, it is worth emphasizing the sensible use of NRI data as far as erosion by wind is concerned.

The NRI confirms the Great Plains as the major area of wind erosion concern, but it should not be used to assess the relative importance of the two forms of erosion where both are relatively minor. Tons of erosion by water cannot be added to tons of soil eroded by wind, and only in a very general way can the relative impact of the two forms of erosion on the potential productivity of U.S. soils be assessed. Still, the NRI can provide evidence that erosion by wind is of major national concern and that more research on this is desperately needed if a complete assessment of the dangers of erosion is ever to be available.

6

Erosion on Range and Forest Lands: Impacts of Land Use and Management Practices

R. Neil Sampson

Range and forestlands account for about 53 percent of the nonfederal lands included in the 1982 National Resources Inventory (NRI) and are estimated to produce 28 percent of the total annual soil loss. Thus, even though this soil loss is less serious than that of intensively cultivated lands, it is an important part of the U.S. soil erosion problem that needs to be addressed. In some areas, the damage occurring to range and, to a much lesser extent, forestlands is the primary resource concern.

This study used the 1982 NRI data to test ways to address the following policy questions:

- How much could future land use decisions affect the soil erosion problem on range and forestlands? For example, if the range and forestland identified as potential cropland were all converted to this use, how would the remaining range and forest be affected? Would the conversion of the best lands leave the remainder significantly more erosion-prone, in terms of the average quality of the resource base?
- What impact will improved range management and timber stand practices have on soil erosion? These practices are generally promoted on the basis of their benefits to the grass or trees, and for improved productivity, but what about the effects on soil erosion? Can areas be identified where more effective range conservation or farm forestry programs can yield important secondary benefits for erosion control?
- Improved targeting of conservation programs has been a much-touted benefit of the broad area information provided by the NRI. How can the 1982 inventory be used to improve targeting on range and forestlands?

To address these policy questions, the study first evaluated the NRI data and tested different analytical methods for effectively using them. In so doing, the following questions were raised:

- How much additional value can be gained from using the Major Land Resource Area (MLRA) data versus state-level or national data? Can substate distinctions be made that will be helpful to conservation policymakers?
- At what point does the data contain too few sample points to be reliable?
- Can a separation of the physical data from the management data in the Universal Soil Loss Equation (USLE) give some insight into the potential for improving erosion conditions through changing management? In the USLE, this would mean letting the RKLS (rainfall, soil erodibility, length of slope, and steepness of slope) indicate the physical condition and using the C (cover) and P (conservation practices) factors to indicate management.

Since the P factor is unity (1) on range and forestland, the C factor becomes the major determinant of the difference between potential erosion (RKLS) and observed erosion (USLE). On rangelands, wind erosion may be more important than sheet and rill erosion in some areas, so the data recorded for the Wind Erosion Equation (WEE) come into play. On forestland, wind erosion can be ignored.

The analysis was conducted in two sections--six MLRAs were chosen for range evaluation and six others for the forestland study. In addition, two state summaries were prepared for both forest and range. (Detailed tables on the characteristics of the areas studied are available from the author.)

A microcomputer spreadsheet program evaluated key characteristics of the MLRA data as a means of helping select MLRAs where certain range and forest factors would be significant. The areas were chosen for a combination of characteristics, including geographic area, percentage of land indicating a particular problem, and total size of the range or forestland resource in the area. Once analytical tables were designed and MLRAs chosen, the Agricultural Research Service (ARS) Soil and Water Management Research Unit at the University of Minnesota did the necessary programming to generate the data tables. The information was transferred to a microcomputer using a spreadsheet program, which allowed manipulation of the basic data. All tables in this paper are the product of the author's computer, not the ARS computer at Minnesota.

Two important caveats about the results for the sample MLRAs and the states need to be mentioned. First, these tables can be easily duplicated by the Minnesota computer or any other mainframe that will run the NRI tapes, but some reprogramming is necessary. Second, the final results of the two methods will not be identical. The Minnesota tables are generated by taking the individual point data and computing weighted averages based on the values at each point. The microcomputer calculations are made by manipulating the weighted average values themselves. Although the results are very similar, they are not identical, and analysts who reproduce this methodology on a mainframe computer using the point data will find minor discrepancies in the results for these sample MLRAs. The methodology, however, should work equally well using either method.

Even though the point sample data are invaluable for certain analytical techniques, much can be learned from the national, state, and MLRA summary data published by the Soil Conservation Service (SCS), without resorting to expensive use of the sample point data. An ordinary microcomputer can aid in ranking MLRAs on the basis of problems or features that are helpful, for example, in evaluating the targeting potential of national programs.

Soil erosion problems can be identified from the tables in this paper and many inferences drawn from them that will be of use to national policymakers. The benefits of changing land use and applying conservation practices are fairly easy to estimate, if certain critical assumptions can be made. But such assumptions need further discussion and agreement within the professional community if they are to provide the basis for future policy research using the NRI data.

RANGELAND

General Findings

Nonfederal rangelands comprise 405.9 million acres, with only about 136 million, or 33.5 percent, rated in the 1982 NRI as adequately protected. Soil erosion was estimated to be at or below the soil loss tolerance limit (T value) on 336 million acres, or 82.8 percent. Erosion rates between T and 2T were associated with 26 million acres (6.4 percent), while almost 44 million acres (10.8 percent) were suffering soil erosion in excess of 2T. Average soil losses on these highly eroding rangelands were estimated at 20.3 tons/acre/year.

The average annual erosion rate on U.S. rangelands is estimated to be about 1.5 tons/acre/year from wind erosion and 1.4 tons/acre/year from sheet and rill erosion. But this 2.9 tons/acre/year average does not reflect some critically eroding areas, which seem to be associated with MLRAs where wind erosion rates are high. In some areas, over 85 percent of the soil loss due to wind erosion is concentrated on less than 1 percent of the land.

That fact alone would seem to provide a strong argument for targeting conservation efforts to those lands. Care must be taken in arriving at that conclusion, however, because much of that soil loss may not be treatable, according to SCS judgments made during the NRI sampling process. A rough estimate of the amount of treatable land can be made from the NRI, however, and this should be highly valuable in designing program targeting efforts.

Range improvements yield significant erosion reductions, even on lands where current erosion problems are not rated as serious. Thus, efforts to improve range conditions through grazing management or to improve or reestablish improved stands of forage are likely to yield significant erosion control benefits in addition to increased grazing, wildlife, and watershed values.

Point sample data from six MLRAs (10, 30, 43, 67, 77, and 81) and from Idaho and Texas were summarized to seek answers to the following questions:

- If the rangeland that was identified as potential cropland were developed, would the remaining rangeland be significantly changed in terms of erosion characteristics? What kind of a conservation problem would exist on the remaining rangeland, assuming it was used at the same intensity and with the same management that it now receives?
- How much erosion control benefit could be achieved by improving the condition of the range forage by the equivalent of one condition class?
- How much erosion control benefit could be achieved by carrying out the forage improvements indicated as needed in the NRI?

● How useful are the 1982 NRI data in identifying areas where it would be beneficial to target range conservation efforts?

Conversion of Rangeland to Cropland

Insights into these questions are relatively easy to obtain from the 1982 NRI data. Table 1 is a summary of data generated by the University of Minnesota, in which weighted averages for total potential erosion (RKLS), total actual sheet and rill erosion (USLE), and total actual wind erosion were calculated for rangeland in each MLRA and in the two states considered. The rangeland point samples with potential for conversion to cropland were then subtracted from the rangeland totals. The result is the erosion characteristics of the rangeland that would remain if the areas with high and medium potential as cropland were actually converted to that new use.

With the percentage increases in total erosion running from 0 to almost 15 percent, it seems clear that there would be very different soil erosion impacts in different regions if the rangeland that could be cropland were converted. The methodology used here seems useful in portraying those regional differences. Although actual range productivity would be difficult, if not impossible, to assess from the NRI data, some of the rangeland characteristics of the land judged to have high and medium potential for conversion to cropland can be evaluated.

Table 2 shows the rangeland conditions of total rangeland, range with cropland potential, and total rangeland minus potential cropland on the sample MLRAs and states. By comparing (with a chi-square test) the condition class distribution found on all rangeland with that found for rangelands that could be cropland, it can be established whether the potential cropland would be taken disproportionately from one or another range condition. This could lead to policy conclusions about the potential damage to the rangeland resource that might occur should the conversions actually take place.

Interesting differences arose among the sample MLRAs and states. In MLRAs 10 and 43, as well as in Idaho, potential cropland would be taken disproportionately from land now listed as in poor range condition. Thus, cropland conversions in this area would have less of an

TABLE 1 Potential and Actual Erosion of Rangeland in Selected MLRAs and States with Percentage Increase of Potential Cropland Converted

Erosion on Rangeland (Tons/Acre)	MLRA 10	30	43	67	77	81	Idaho	Texas
Average potential erosion (RKLS)								
All rangeland	21.84	2.93	44.32	14.69	9.57	24.85	36.57	19.91
Potential cropland	16.65	0.96	11.15	7.44	5.48	7.92	13.60	13.04
Rangeland without potential cropland	22.09	3.07	45.82	16.54	10.49	25.89	38.42	21.07
Average actual sheet/rill erosion (USLE)								
All rangeland	1.45	0.59	1.31	1.21	0.81	1.59	0.55	1.21
Potential cropland	0.93	0.21	0.23	0.53	0.39	0.37	0.31	0.59
Rangeland without potential cropland	1.47	.62	1.36	1.38	0.90	1.67	0.57	1.32
Average actual wind erosion (WEE)								
All rangeland	.00	41.27	.00	0.40	3.67	0.13	0.04	0.65
Potential cropland	.00	6.88	.00	0.99	1.40	0.01	0.11	0.19
Rangeland without potential cropland	.00	43.76	.00	0.24	4.19	0.13	0.03	0.72
Average total erosion (USLE + WEE)								
All rangeland	1.45	41.86	1.31	1.60	4.48	1.72	0.59	1.86
Potential cropland	0.93	7.09	0.23	1.53	1.80	0.37	0.43	0.78
Rangeland without potential cropland	1.47	44.38	1.36	1.62	5.09	1.80	0.60	2.04
If potential cropland were converted:								
Percent increase in RKLS	1.14	4.78	3.38	12.59	9.61	4.19	5.06	5.83
Percent increase in USLE	1.38	5.08	3.82	14.05	11.11	5.03	3.64	9.09
Percent increase in WEE	.00	6.03	.00	-40.00	14.17	0.00	-25.00	10.77
Percent increase in USLE + WEE	1.38	6.02	3.82	01.25	13.62	4.65	1.69	9.68

SOURCE: Derived from 1982 NRI.

TABLE 2 Percentage of Rangeland, Selected MLRAs and States in Various Forage Conditions and with Conservation Needs

Rangeland Conditions	MLRA 10	30	43	67	77	81	Idaho	Texas
Excellent condition class								
All rangeland	2.72	8.35	7.46	1.33	0.60	0.56	5.10	0.51
Potential cropland	.00	5.37	2.60	1.87	0.63	0.44	0.37	0.52
Rangeland without potential cropland	2.84	8.70	7.67	1.19	0.60	0.56	5.40	0.51
Good condition class								
All rangeland	20.14	20.45	32.30	30.53	30.58	7.26	34.55	14.53
Potential cropland	5.00	39.69[a]	26.79	29.51	38.10	4.75	28.57	14.54
Rangeland without potential cropland	20.81	18.19	32.54	30.80	28.90	7.42	34.93	14.52
Fair condition class								
All rangeland	34.21	47.08	45.88	58.50	57.76	73.56	40.53	57.42
Potential cropland	29.11	45.15	46.13	59.21	48.96	85.23	36.45	56.52
Rangeland without potential cropland	34.43	47.31	45.87	58.32	59.71	72.85	40.79	57.57
Poor condition class								
All rangeland	42.92	24.12	14.35	9.63	11.06	18.62	19.83	27.54
Potential cropland	65.89[a]	9.79	24.49[a]	9.41	12.30	9.57	34.61[a]	28.42
Rangeland without potential cropland	41.92	25.81	13.93	9.69	10.79	19.17	18.88	27.39
Adequately protected								
All rangeland	15.17	19.56	30.65	33.97	35.87	15.73	22.35	22.84
Potential cropland	11.73	42.48	36.08	41.07	46.15	24.81	24.60	29.16
Rangeland without potential cropland	14.64	16.69	29.08	25.61	27.37	14.29	20.52	18.63
Needs forage improvement								
All rangeland	43.07	8.94	32.86	26.17	35.92	65.30	39.60	51.56
Potential cropland	55.72	5.22	20.55	28.88	30.61	67.02	38.52	46.12
Rangeland without potential cropland	40.54	8.59	31.97	20.29	30.29	61.42	36.74	44.90
Needs forage reestablishment								
All rangeland	14.29	2.77	4.25	2.25	3.80	9.43	11.36	13.65
Potential cropland	26.24	.00	11.22	2.46	4.96	5.35	27.85	17.28
Rangeland without potential cropland	13.10	2.77	3.77	1.75	2.88	9.12	9.29	11.15

[a]Significant at the 95 percent level.

SOURCE: Derived from 1982 NRI.

effect on the average quality of the remaining rangeland than might be the case in MLRA 30, where there appears to be a tendency for potential cropland to be in good range condition. In MLRA 67 and in Texas, the condition of the land shown as potential cropland is almost identical to the average condition of the total range resource. In MLRAs 77 and 81, the potential cropland seems to fall somewhat disproportionately in the good and fair condition categories, but the differences are not significant at the 95 percent level.

In all but MLRA 10, it appears that the potential cropland would be disproportionately drawn from rangeland that is now adequately protected. In MLRA 10, potential cropland would come disproportionately from lands needing forage improvement or reestablishment, which seems consistent with the fact that these same lands were rated as being in poor range condition.

Erosion Control Potential in Range Management

Range conservationists promote improved range management for many reasons, including soil conservation. Ranchers are primarily interested in the potential for improved forage production through management changes. Thus, most of the literature promoting range conservation focuses on the potential improvements in grazing values that can be gained by improving range conditions. But policymakers must look at range conservation programs to determine the overall public benefits. If significant reductions in soil erosion are part of the benefits, range programs have one more bargaining point as they compete for national and state funds.

The 1982 NRI offers an excellent opportunity to estimate the soil erosion control that might result from improvements in the forage condition on rangelands. The methodology is relatively straightforward. Table 3 summarizes the NRI point sample data for MLRA 77, as an example, to calculate the acreage in each rangeland condition class, plus the weighted average of the factors. Similarly, Table 4 addresses wind erosion. These were easily converted into estimates of current soil loss in tons. (Tables 3 and 4 are in 100 tons because the acreage figures were generated in 100-acre units.)

At this point, some assumptions were made. If the rangeland that is now in good condition was improved to

TABLE 3 Estimated Sheet and Rill Erosion Reduction in MLRA 77 from Improving the Condition of Perennial Range Vegetation by One Condition Class

Rangeland Condition Class	Acreage (100 Acres)	Weighted Average				Soil Loss (100 Tons)	Improved C Factor	Potential Erosion Reduction	
		RKLS	C	P	USLE			(100 Tons)	Percent
Excellent	835	7.50	.087	1	.70	545	.09	0	.00
Good	42,238	9.40	.064	1	.64	25,410	.06	0	.00
Fair	79,783	10.00	.084	1	.90	67,018	.06	15,957	23.81
Poor	15,283	7.80	.106	1	.82	12,636	.08	2,623	20.75
Annual range	--	--	--	--	--	0	--	0	--
Not applicable	622	4.50	.100	1	.42	280	.100	0	.00
Not rangeland	--	--	--	--	--	0	--	0	--
Total	138,761					105,889		18,579	17.55

SOURCE: Derived from 1982 NRI.

TABLE 4 Estimated Wind Erosion Reduction in MLRA 77 from Improving the Condition of Perennial Range Vegetation by One Condition Class

Weighted Average WEE	Soil Loss (100 Tons)	Improved WEE	Potential Erosion Reduction (100 Tons)	Percent
.00	0	.00	0	.00
2.75	116,155	.00	116,155	100.00
3.20	255,306	2.75	35,902	14.06
8.89	135,866	3.20	86,960	64.00
0	0	.00	0	.00
4.06	2,525	4.06	0	.00
Total	509,851		239,017	46.88
Total wind + sheet and rill erosion	615,740		257,596	41.84

SOURCE: Derived from 1982 NRI.

excellent condition, the weighted average C factor associated with excellent condition in that MLRA should also be achieved. Similarly, a move from fair to good condition should improve the C factor accordingly, as would a move from poor to fair condition. This would work very well, except where the weighted average C factor is actually lower (indicating better cover conditions) on the poorer condition range.

In MLRA 77, for example, the C factor on good condition range is lower than that recorded on excellent condition range. When that occurs, a shift from good to excellent condition is not assumed to be accompanied by an increase in C factor, so the C factor now existing on the good condition rangeland was used.

The column labeled "Improved C Factor" thus takes either the C factor of the next higher condition class or the existing C factor, whichever is lower. It was a relatively simple task, then, to multiply the improved C factor by the RKLS for the land in question and to estimate the potential erosion reduction that might be achieved from moving up one condition class (and simultaneously achieving the new C factor).

Wind erosion requires a slightly different methodology, since only the WEE final estimate, not the individual equation factors, are contained in the NRI point data. So it was assumed that the achievement of an improved rangeland condition class would result in the same WEE estimate as already experienced by other lands in that higher class in the MLRA (see Table 4). An exception was made where the WEE was not inversely related to range condition class, in which case it was assumed that improving range condition would not induce more wind erosion, but that the WEE would remain the same.

Table 5 shows a summary of the results obtained in the six MLRAs and two states tested with this method. On the basis of this sample, it appears that the erosion control benefits of range improvement might be significant--causing a drop of approximately 31 percent in combined wind and water erosion on rangelands. The potential benefits vary widely, however, so analysis of the NRI data could be very useful in identifying both the regions where potential benefits might be the highest, as well as the type of erosion (wind or water) that might be most affected by improved range management.

An interesting variation of this method tested the Resource Conservation Act (RCA) goal currently being proposed by SCS--to raise the condition of poor and fair rangelands to good condition. Table 6 shows how this might affect erosion. The methodology followed was exactly the same as for Tables 3, 4, and 5, but both excellent and good condition ranges were left unchanged, while poor and fair condition ranges were adjusted so that the C factor and WEE products either equaled the current C and WEE associated with good condition land or were unchanged if they were already in a less erosion-prone condition.

The effect of achieving this goal would be somewhat different than that of achieving a one-class improvement across the board. For the MLRAs tested, the total soil erosion reduction would be slightly less (dropping from 30.74 to 27.95 percent). The major difference appears to be that the RCA goal might be slightly more effective in reducing sheet and rill erosion, and slightly less effective in reducing wind erosion. In some MLRAs (10, for example) the RCA goal seems to hold more promise for erosion control, while in others (77, for example) it

TABLE 5 Estimated Erosion Reduction from Improving the Condition of Perennial Range Vegetation by One Condition Class, Selected MLRAs and States (Percentages)

Erosion Type	MLRA 10	30	43	67	77	81	Idaho	Texas	Average
Sheet and rill erosion	29.25	2.31	20.18	22.04	17.55	43.83	15.74	39.72	23.83
Wind erosion	.00	23.22	.00	77.59	46.88	26.32	75.67	34.20	35.49
Total erosion	29.25	22.85	20.18	34.52	41.84	42.56	18.44	36.29	30.74

SOURCE: Derived from 1982 NRI.

TABLE 6 Estimated Erosion Reduction from Improving Range in Poor and Fair Condition to Good Condition, Selected MLRAs and States (Percentages)

Erosion Type	MLRA 10	30	43	67	77	81	Idaho	Texas	Average
Sheet and rill erosion	46.30	.00	15.78	17.44	19.80	51.87	7.32	46.92	25.68
Wind erosion	.00	23.59	.00	82.23	25.45	16.37	80.81	29.08	32.19
Total erosion	46.30	23.17	15.78	17.62	24.48	49.30	11.09	35.83	27.95

SOURCE: Derived from 1982 NRI.

seems to hold considerably less. This may argue for a more flexible goal-setting process, done on a statewide, rather than national, basis.

Erosion Control Potential in Rangeland Improvement Practices

The NRI data can also be used to estimate the potential erosion control benefits of applying the conservation practices listed as needed on rangelands. To estimate the soil loss reductions that might be achieved, the point data on acres needing various treatments, by MLRA, were used to prepare a table containing the weighted averages of the RKLS, USLE, and WEE factors. (The C factor is also needed. In the tables generated for this paper, it was not obtained by summing from the point samples, so it had to be deduced by dividing the weighted average of USLE by the weighted average of RKLS.)

For each treatment need, it was assumed that applying the needed treatment would achieve a C factor equal to that of the adequately treated land in the MLRA or state. Therefore, the new erosion level was calculated by multiplying the best attainable C factor by the existing RKLS, and the percent erosion reduction was calculated.

Table 7 summarizes the findings for all six MLRAs and for Idaho and Texas. In looking at this table, it is important to realize that the percentage erosion reductions apply only to those acres shown as needing each individual treatment; they are not applicable to the entire range resource of the area in question.

One MLRA, however, has enough unusual characteristics to raise questions about whether this methodology should be attempted there without considerable further investigation. In MLRA 30, the Sonoran Desert, there are very high wind erosion rates shown on much of the rangeland. USLE estimates are very low, as would be expected with the very low R factor (rainfall) involved. Nearly half the land and 65 percent of the soil loss in MLRA 30 are rated as unfeasible to treat.

What is even more perplexing, however, is that the weighted annual average wind erosion is over 10 tons/acre/year on land that has been rated as adequately protected. This raises questions about the accuracy of the Wind Erosion Equation as it applies to those desert conditions, or about the judgment of the field people in determining adequate protection, or both. With almost 1 million

TABLE 7 Estimated Erosion Reduction from Applying Needed Conservation Treatments on Rangelands, Selected MLRAs and States (Percentages)

Conservation	MLRA							
Treatment	10	30	43	67	77	81	Idaho	Texas
Erosion control	73.63	79.1	81.95	55.32	92.37	81.01	74.92	86.05
Forage protection only	42.31	.00	59.83	36.77	83.89	66.07	32.95	67.21
Improvement without brush	67.14	88.05	67.9	34.04	78.79	70.94	53.69	75.39
Improvement with brush	76.48	87.49	70.88	58.05	52.66	68.12	48.99	65.68
Reestablish without brush	77.84	53.57	75.21	93.20	92.04	74.11	65.50	78.92
Reestablish with brush	85.15	1.91	20.18	71.42	32.79	82.63	73.85	73.94
Treatment unfeasible								
Percent land	9.12	48.66	3.82	.38	1.1	2.26	5.82	2.67
Percent erosion	4.66	65.34	21.75	1.93	4.38	7.67	6.29	16.33

SOURCE: Derived from 1982 NRI.

acres in the sample category, the problem does not appear to be caused by a few renegade sample points, so additional checking or explanation seems in order.

Using the 1982 NRI Data to Target Range Conservation Programs

A great deal of the soil erosion problem is concentrated on a small percentage of the land, which has been cited in recent USDA efforts to target soil conservation programs toward those lands where technical assistance, cost-sharing, or other forms of conservation incentive can produce the greatest reduction in soil loss per federal dollar. As in 1977, the 1982 NRI confirms this pattern. Point data for the sample MLRAs and states were aggregated according to ranges of actual soil loss, as indicated by the USLE and WEE results, and a weighted average soil loss was generated for each aggregation. Multiplying this weighted average by the acreage in the category gave total tons of soil loss, which enabled cumulative percentages for both acreage and soil loss to be generated.

Table 8 shows the results of this method for sheet and rill erosion in MLRA 43, and Table 9 provides wind erosion data from MLRA 77. In both these examples, a good case can be made for targeting soil conservation work on rangelands. In MLRA 43, about half the sheet and rill erosion is occurring on only about 5 percent of the land--those areas that are eroding at rates of 5 tons/acre/year or higher. In MLRA 77, 96 percent of the wind erosion is occurring on only about 20 percent of the land--areas eroding at rates of 2 tons/acre/year or higher. MLRA 30, the Sonoran Desert, demonstrated a somewhat different distribution, as might be expected, with 62 percent of the land suffering no wind erosion at all, while 16 percent was losing almost 85 percent of the total soil loss attributed to wind erosion n the MLRA.

That would be a strong case for targeting, but it should be tempered by remembering that 49 percent of the land and 65 percent of the soil loss was rated as unfeasible to treat in this MLRA. This no doubt means that natural geologic erosion rates are high and that USDA program targeting would have little, if any, effect.

(For the tests run in this assessment, the University of Minnesota ran the special computer program that counts the number of point samples in each segment of each

TABLE 8 Estimated Sheet and Rill Erosion on Rangeland, MLRA 43

Actual Erosion (USLE)	Acreage (100 Acres)	Cumulative Percent of Acreage	Weighted Average Sheet/Rill Erosion	Soil Loss (100 Tons)	Cumulative Percent of Erosion
0--<1	50,624	73.85	0.25	12,656	14.05
1--<2	8,060	85.61	1.45	11,687	27.02
2--<3	3,738	91.06	2.50	9,345	37.40
3--<4	1,353	93.04	3.42	4,627	42.53
4--<5	1,106	94.65	4.40	4,866	47.94
5--<6	479	95.35	5.50	2,635	50.86
6--<7	610	96.24	6.45	3,935	55.23
7--<8	275	96.64	7.27	1,999	57.45
8--<9	111	96.80	8.72	968	58.52
9--<10	185	97.07	9.47	1,752	60.47
10--<11	83	97.19	10.39	862	61.43
11--<12	124	97.37	11.33	1,405	62.99
12--<13	129	97.56	12.28	1,584	64.74
13--<14	948	98.95	13.33	12,637	78.77
14--<15	0	98.95		0	78.77
15--<20	372	99.49	16.85	6,268	85.73
20--<25	25	99.52	23.00	575	86.37
25--<30	5	99.53	29.36	147	86.53
30--<50	294	99.96	36.37	10,693	98.40
50--<75	27	100.00	53.28	1,439	100.00
75--<100	0	100.00		0	100.00
100 & up	0	100.00		0	100.00
Total	68,548			90,079	

SOURCE: Derived from 1982 NRI.

category in the tables. In the MLRAs and states chosen, the rangeland acreages were fairly large, so the lack of point samples to ensure data accuracy did not seem to present a problem. Even the smaller acreage divisions were the result of 5 to 10 sample points, and the assumption is that anything represented by four or more points is a statistically reliable number for these purposes. One exception was in the soil erosion groupings used for MLRAs 43 and 77, where some of the categories lacked adequate point data. One way to get around this problem, with little apparent loss of utility to the analysis, would be to group the erosion levels in broader groups. For rangeland, it would appear that groupings of less than 1 ton/acre/year, 1 to 2, 2 to 5, 5 to 10, 10 to 25, 25 to 50, and over 50 would be adequate for most evaluations and would ensure that adequate point samples existed in each category so that statistical reliability could be maintained.)

Several additional evaluations could make this analysis more relevant:

● Sample point data could be summarized to aggregate only those points rated as feasible to treat, which would be a much more logical basis for comparing different MLRAs to see where targeting would be the most productive.

● NRI data were collected on the trends in rangeland condition (up, even, or down) and the estimated grazing level at the time of the sampling (not grazed, presently deferred, properly used, or excessively used). The point data were summarized according to these factors, and associated RKLS, C, P, USLE, and WEE weighted averages obtained. It would be interesting to correlate the C factors obtained under proper grazing use with those associated with excessive use to yield an estimate of the erosion-prevention value of promoting proper grazing use on rangeland.

● Comparisons with the individual soil characteristics contained in Soils-5 can be made to separate the sample point data by total soil erosion rates compared with the soil loss tolerance limit. There was not enough time to do this, but a computer run could be designed to separate the acreage shown as needing the various

TABLE 9 Estimated Wind Erosion on Rangeland, MLRA 77

Actual Erosion (WEE)	Acreage (100 Acres)	Cumulative Percent of Acreage	Weighted Average Wind Erosion	Soil Loss (100 Tons)	Cumulative Percent Erosion
0--<1	103,293	74.44	0.08	8,263	1.62
1--<2	8,974	80.91	1.41	12,653	4.10
2--<3	6,277	85.43	2.49	15,630	7.17
3--<4	2,239	87.04	3.36	7,523	8.65
4--<5	1,967	88.46	4.44	8,733	10.36
5--<6	1,373	89.45	5.38	7,387	11.81
6--<7	1,596	90.60	6.48	10,342	13.84
7--<8	919	91.26	7.48	6,874	15.19
8--<9	441	91.58	8.43	3,718	15.92
9--<10	695	92.08	9.47	6,582	17.21
10--<11	310	92.31	10.52	3,261	17.85
11--<12	535	92.69	11.46	6,131	19.05
12--<13	868	93.32	12.48	10,833	21.18
13--<14	503	93.68	13.48	6,780	22.51
14--<15	466	94.01	14.75	6,874	23.86
15--<20	1,667	95.22	16.76	27,939	29.34
20--<25	1,610	96.38	21.81	35,114	36.23
25--<30	1,246	97.27	27.53	34,302	42.96
30--<50	1,721	98.51	38.54	66,327	55.98
50--<75	543	98.91	58.13	31,565	62.17
75--<100	446	99.23	86.29	38,485	69.73
100 & up	1,072	100.00	143.91	154,272	100.00
Total	138,761			509,588	

SOURCE: Derived from 1982 NRI.

conservation treatments into three categories: land eroding at less than T, land eroding at T to 2T, and land eroding at over 2T. Such a table would be useful in assessing where the application of these conservation measures could help treat the most vulnerable rangeland soils.

FORESTLAND

General Findings

The 1982 NRI identified 393.7 million acres of nonfederal forestland, some 26 percent of the total nonfederal estate. About 37 percent of this land was adequately protected, with timber stand treatment needed on most of the rest. Soil erosion rates at or below the established T value were found on 94 percent of the forestland, with another 9.8 million acres (2.5 percent) eroding at levels between T and 2T and 13.6 million acres (3.5 percent) eroding at over 2T. Clearly, soil erosion is not a serious problem on most forestland.

Where erosion exists, however, it can be intense, because forestlands are generally steeper than more intensively used lands, with topsoils that are often thin and vulnerable to damage. Consequently, even though the average annual erosion rate on forestlands is just under 1 ton/acre/year, there are places where erosion control is badly needed.

Much of the accelerated erosion on forestland is associated with the grazing of livestock, and the division of the NRI statistics into grazed and ungrazed forestland helps assess this important difference. Soil erosion, on the national average, is about four times more severe on grazed forestland than on ungrazed. In some MLRAs (126, for example), grazed forestland is very erosion-prone, with RKLS factors in the 400 to 600 range. Such land is so steep and susceptible to erosion that removing livestock may be the best (or only) way to treat the erosion problem.

In contrast to the findings on rangeland, most soil erosion on forestlands can be treated. Even where erosion rates were highest, the sample NRI data tested suggest that very little of the problem is not feasible to treat, and the application of needed forest management and timber stand improvement practices appears to have the potential of reducing soil erosion by one-third to three-fourths.

Point sample data from six MLRAs (1, 15, 105, 115, 126, and 133B) and from Georgia and Michigan were summarized to consider the following questions:

- If potential cropland were developed, would the average quality of the remaining forestland be significantly changed in terms of erosion characteristics or timber productivity?
- How much erosion control benefit could be achieved by carrying out the erosion control and timber stand improvements shown in the NRI as needed?
- How can the 1982 NRI be used to improve understanding of the nation's nonindustrial private forestlands, which offer a significant challenge to forestry programs in many agencies and organizations?

Conversion of Forestland to Cropland

A great deal of excellent forestland has been lost to crop production in recent years, and more may be lost in the future. One question that can be evaluated easily with the NRI data deals with the impact of future conversion on the soil erosion problem of remaining forestland.

Point sample data were summarized by grazed and ungrazed forestland, and a weighted average developed for both RKLS and USLE. The points shown as having high or medium potential for conversion to cropland were identified, and similar weighted averages developed for them, thus providing an estimate of the erosion characteristics on the remaining forestland should those lands actually be converted. Table 10 summarizes the results from the six MLRAs and two states tested.

Several things are evident from this table. First, in most areas and on both grazed and ungrazed forest, the comparative RKLS factors show that the land with potential for cropland is much less erosion-prone than the average for all forestland. Removing the better land is going to leave the remaining forest somewhat more erosion-prone, on the average, than before. There is significant variation among MLRAs and states on this, however. In MLRA 15, for example, the effect would be negligible, while in MLRA 115, farmland conversion could leave the remaining forestland as much as 10 percent more erosion-prone than it is today. If that occurred, a region where forestland might not appear to have a soil erosion problem today might indeed have such a problem in the future.

TABLE 10 Potential and Actual Erosion of Forestland in Selected MLRAS and States, with Percentage Increase of Potential Cropland Converted

Average Erosion on Forestland (Tons/Acre)	MLRA 1	15	105	115	126	133B	Georgia	Michigan
Potential erosion (RKLS)								
All ungrazed forestland	288.11	181.23	373.83	269.99	474.85	39.40	94.02	16.21
Potential cropland	72.94	109.65	98.71	53.98	154.95	21.40	30.32	6.47
Ungrazed forestland without potential cropland	309.58	183.12	402.21	299.38	500.00	40.49	104.91	17.37
Actual sheet/rill erosion (USLE)								
All ungrazed forestland	2.73	8.64	0.99	1.59	1.50	0.38	0.23	0.14
Potential cropland	0.08	6.88	0.37	0.25	0.25	0.15	0.08	0.13
Ungrazed forestland without potential cropland	2.99	8.69	1.06	1.77	1.60	0.40	0.25	0.14
Potential erosion (RKLS)								
All grazed forestland	239.10	135.46	452.67	326.96	555.40	47.47	119.03	18.83
Potential cropland	13.33	91.80	137.52	100.13	163.87	23.11	53.22	7.61
Grazed forestland without potential cropland	248.85	135.80	486.68	354.81	583.94	50.55	133.44	25.78

Actual sheet/rill erosion (USLE)								
All grazed forestland	3.07	7.49	5.10	13.04	7.65	0.99	0.73	0.62
Potential cropland	0.04	4.69	1.72	3.11	1.68	0.41	0.30	0.10
Grazed forestland without potential cropland	3.20	7.51	5.46	14.25	8.09	1.06	0.83	0.93
If potential cropland were converted:								
Ungrazed forestland								
Percent increase in RKLS	7.45	1.04	7.59	10.89	5.30	2.77	11.58	7.16
Percent increase in USLE	9.52	0.58	7.07	11.32	6.67	5.26	8.70	0.00
Grazed forestland								
Percent increase in RKLS	4.08	0.25	7.51	8.52	5.14	6.49	12.11	36.91
Percent increase in USLE	4.23	0.27	7.06	9.28	5.75	7.07	13.70	50.00

SOURCE: Derived from 1982 NRI.

One interesting aspect of the 1982 NRI was the attempt to identify general forest types during the sampling. This provides an opportunity to look at the kinds of forests that seem most susceptible to continued conversion to cropland. Table 11 shows the distribution of the general forest types within the test areas, and Table 12 indicates the percentage of each type in each test area that might be converted to cropland.

It is not uncommon in these sample areas for 10 to 25 percent of a given forest type to be rated as having potential for conversion to cropland. Just what impact this would have on the forest products industry in those regions is beyond the scope of this study, but it would appear that several policy inferences could be made from these data, particularly if they were analyzed on a state-by-state basis, using MLRA data, within each state to identify regional impact potentials. In the samples from the central and eastern regions, the large acreages of oak-hickory forest that might be converted to cropland seem to hold the largest potential impact.

Another way to use the NRI in looking at potential land use impacts is to summarize the effects of potential cropland conversion on the remaining forest resource as indicated by the current canopy cover of the forestland most likely to be converted. This gives some idea of the size and value of the forest stands on those lands. Tables 13 and 14 are calculated in exactly the same manner as the two preceding tables, with the acreage figures aggregated according to estimated forest canopy cover. As can be seen, most of the potential cropland has canopy covers of less than 50 percent, and, in many areas, one-quarter to one-third is associated with a canopy cover of less than 25 percent. In some areas, the distribution is almost equally split among the canopy cover categories. Of interest is MLRA 1, where most of the forest has a canopy cover of over 50 percent, and almost 11 percent of that forestland is rated as having potential for conversion to cropland.

The NRI forestland data also captured information about the size of trees, separating those with a diameter at breast height (DBH) of over 5 inches from the smaller trees. On areas with average DBH of less than 5 inches, a stocking rate was estimated. Those factors were not correlated with conversion potential in this analysis, but that could be done if it were seen as potentially useful.

TABLE 11 General Forest Types in Selected MLRAs and States (100 Acres)

General Forest Type	MLRA							
	1	15	105	115	126	133B	Georgia	Michigan
Jack pine			835	216	1,942		184	16,928
Spruce/fir			67		51		12[a]	24,916
Loblolly/slash pine					12[a]	6,794	43,050	
Loblolly/shortleaf pine			9[a]	173	2,176	71,543	45,891	10[a]
Oak/pine			256	822	1,953	80,718	64,422	3,280
Oak/hickory		10[a]	25,556	28,685	57,269	20,940	42,184	16,309
Oak/gus/cypress				694	59	11,040	16,896	
Elm/ash/cottonwood			1,846	4,053	1,729	558	413	13,463
Maple/beech/birch			860	197	6,281		59[a]	40,484
Aspen birch			1,073	19[a]	353			35,462
Low production type		8[a]	11[a]	25[a]	2,042	48[a]	131	
Nonstocked	1,609	67[a]	199	309	1,126	5,938	5,594	2,727
Douglas fir	33,831	226						18[a]
Ponderosa pine		670						
Fir/spruce	175							
Hemlock/sitka spruce	12,707							
Lodgepole pine	155							
Redwood	26[a]	308						
Hardwoods	9,602	16,591		12[a]				
Other conifers		32[a]						
Savanna		10,167						
Total	58,105	28,079	30,712	35,205	74,993	197,579	218,836	153,597

[a]Data cells of dubious statistical value--four sample points or fewer.

SOURCE: Derived from 1982 NRI.

TABLE 12 Percentage of Land by General Forest Types, with Potential for Conversion to Cropland, Selected MLRAs and States[a]

General Forest Type	MLRA 1	15	105	115	126	133B	Georgia	Michigan
Jack pine			6.47		14.37			8.69
Spruce/fir								4.16
Loblolly/slash pine						20.36	25.43	
Loblolly/shortleaf pine					12.64	5.56	12.46	
Oak/pine			12.89	3.16	6.71	6.43	14.71	6.59
Oak/hickory			9.14	10.04	6.54	8.62	9.69	14.92
Oak/gus/cypress				33.00		3.30	1.78	
Elm/ash/cottonwood			11.97	18.33	9.66	19.35		16.62
Maple/beech/birch			18.49	41.62	10.51			10.74
Aspen birch			8.39		7.65			13.54
Low production type					4.75			
Nonstocked				37.22	3.29	15.96	28.10	16.72
Douglas fir	6.90							
Ponderosa pine		16.12						
Fir/spruce								
Hemlock/sitka spruce	18.60							
Lodgepole pine								
Redwood								
Hardwoods	5.41	1.28						
Other conifers								
Savanna		0.90						
Total	9.02	1.46	9.47	11.78	7.24	6.99	14.68	11.09

[a]Percentages only calculated where the acreage of a given general forest type with potential for conversion to cropland was 2,500 acres or more. Smaller acreages were discarded as having too few sample points for statistical reliability.

SOURCE: Derived from 1982 NRI.

TABLE 13 Forest Canopy Cover in Selected MLRAs and States (100 Acres)

Estimated Canopy Cover (Percent)	MLRA 1	15	105	115	126	133B	Georgia	Michigan
0--9	6,755	10,004	878	481	1,323	12,217	13,520	7,461
10--25	6,300	4,422	1,315	1,435	3,520	11,091	11,438	11,331
26--55	8,583	5,417	4,501	6,892	11,669	40,277	36,888	30,562
56--100	36,467	8,236	24,018	26,397	58,481	133,994	156,990	104,243
Total	58,105	28,079	30,712	35,205	74,993	197,579	218,836	153,597

SOURCE: Derived from 1982 NRI.

TABLE 14 Percentage of Land, by Forest Canopy Cover, That Has Potential for Conversion to Cropland, Selected MLRAs and States

Estimated Canopy Cover (Percent)	MLRA							
	1	15	105	115	126	133B	Georgia	Michigan
0--9	5.67	1.21	10.48	22.04	9.45	6.21	18.56	15.69
10--25	4.49	1.11	5.48	13.52	10.14	8.16	23.01	13.81
26--55	7.18	1.59	11.60	13.31	7.12	7.20	17.24	12.12
56--100	10.86	1.88	9.25	11.10	7.04	6.89	13.14	10.16
Total	9.02	1.46	9.47	11.78	7.24	6.99	14.68	11.09

SOURCE: Derived from 1982 NRI.

Erosion Reduction Potential in Conservation Treatment on Forestlands

In assessing the potential for erosion reduction associated with the application of conservation practices, the NRI point sample data were summarized to develop a weighted average for the RKLS and C factors associated with the various treatment categories for each sample area. The P factor was left out of the tables because it is unity (1) in all cases, and the wind erosion (WEE) was ignored because it was 0 in virtually all cases. It was assumed that the application of the needed treatment would result either in a C (cover) factor similar to that of land adequately protected in the same area, or in the maintenance of the existing C factor, whichever was lower. The analysis was done separately for grazed and ungrazed forestlands. (Detailed tables with the results of this analysis for the six MLRAs and the two states are available from the author.)

Several conclusions emerged. Forestland that is adequately treated has very low soil erosion rates, and the treatment of forestland identified as needing erosion control, if that treatment could achieve good forest cover, has the potential of reducing erosion on those lands by 60 to 90 percent.

Even timber stand improvement, however, may be associated with significant soil erosion reductions (in terms of percentage) if the assumptions are correct. Reductions in the range of 50 percent are not uncommon if an improved cover condition equivalent to that experienced on adequately protected land in the region can be achieved as part of the timber stand work. Since most of the work on improving timber crops includes some attention to roads, trails, and other openings in the forest, this may be achievable.

On the national level, about 9 percent of the forestland needs timber establishment and reinforcement, 42 percent needs timber stand improvement, and 2 percent needs timber crop improvement. If significant progress on these treatment needs could be made, soil erosion on the nation's forests would be virtually nonexistent.

To get at the soil erosion problem, however, conservation programs need only target about 3 to 5 percent of the forestlands in most areas. The 1982 NRI data provide excellent information for locating the general region and size of the areas that need to be targeted, and they give good guidance as to whether the program could be enhanced by attention to timber stand improvement practices or whether it would be best to simply concentrate on erosion control efforts. It should be noted that SCS field staff were instructed to identify land that needed grazing eliminated in order to control erosion as land "needing erosion control." Thus, some of the land in the category of grazed forestland can be assumed to be land where livestock grazing is incompatible with the resource situation. Just how much of the grazed forest is in this condition cannot be determined from the NRI data, however.

Using the 1982 NRI to Evaluate Forestry Program Potential

As the first national statistical sample to include detailed forestry information, the 1982 NRI is of definite interest to the forestry policy community. Although industrial and nonfederal publicly owned forestlands are not separable from nonindustrial privately owned lands, it appears that the data can be of considerable value when used in conjunction with other sources of forest information.

A caveat is necessary, however, based on indicators from the limited sample data reviewed to date. The SCS technicians who were filling out the sample point data for the NRI were not all foresters, and this was a first attempt, so the data on specific forest types and even on general forest types may be somewhat suspect. In this limited analysis, many data cells were encountered where the acreage suggested only one or two points in the entire MLRA.

In assessing the size of the trees on the sample site (DBH) and the stocking rate, the field technician was supposed to list the DBH in inches if it was over 5, to estimate the stocking rate (poor, moderate, full, or nonstock) if the DBH was under 5 inches. In our test MLRAs, there were many samples where neither a DBH nor a stocking rate was recorded. Those points were counted, and from them, a rough estimate of an error rate can be determined.

Whether all of these problems are errors or simply anomolies could not be ascertained, but it was clear that there were reasons to use the forest data with caution. As analyses continue, however, the errors will probably come to light, and the SCS will be able to improve techniques in forestry monitoring on a future NRI. Both the

1982 data and any future versions should be very carefully analyzed, and checked against other data, for value in providing background information and guidance to the nation's state and private forestry programs.

One conclusion could be drawn from this review, however. The forestland data are probably not too important in most states as a source of information on soil erosion and the effects of various conservation treatments. Soil erosion is simply not much of a problem on most forestlands.

This is not to say, however, that the NRI data cannot be of considerable value in the analysis of forestry policy. Tests to develop analytical techniques for evaluating the general forest type, canopy cover, and stand size information should be conducted to test the value of the MLRA and statewide estimates as indicators of the potential workload for forestry programs on nonfederal lands. It appears possible, from this limited review of the data, to use this information as one basis for identifying the forest opportunities of the nation and for drawing some conclusions about where the payoff of targeted forestry programs would be highest.

GENERAL CONCLUSIONS

If the primary goal of conservation programs is the reduction of soil erosion, the point source data contained in the NRI data files can be very helpful in analyzing the areas where targeting effort might be most promising. In addition, particularly for rangeland, they can be used to give some approximations of the returns that might be associated with different targeting schemes.

If four sample points are considered useful criteria for selecting those data elements that have adequate statistical reliability, there seems to be little problem in utilizing the data base where acreages of range and forestland are fairly large. In MLRAs where these acreages are fairly small, however, statistical reliability will be a significant problem. One response would be to add a line on every table generated that would be labeled "all other" or something similar, where the sample units containing less than four sample points could be aggregated. Such a category would allow each table to accurately add the acreages in the MLRA and to keep the internal percentages accurate without misleading the analyst by indicating small amounts of a condition that may or may not exist.

The use of MLRA data rather than state data when using the SCS summary tables seems fully justified. In both forest and rangeland, the MLRA data would lead to different, and more accurately targeted, policy decisions than the statewide averages would. In the rangeland analysis, for example, both MLRA 10 and 43 are located partially in Idaho, and both 77 and 81 are located mainly in Texas.

In both cases, however, the MLRA data were different from each other and from the state data. Thus, it would appear that most of the analyses that would be most useful for program managers could be run as a state analysis using those portions of the MLRA within the state. In many cases, this would lead to rather small areas, and increased problems with statistical reliability, but the use of the "all other" category suggested above should help that situation.

In performing national program and policy tests, where each MLRA can be used in whole for evaluation, it seems clear that MLRA data will be far more useful, particularly when working with the SCS-generated summaries.

Finally, while there are many ways in which the point data are invaluable for research purposes, there are also very useful interpretations that can be made from the summary data provided by SCS. In selecting the MLRAs for this study, a common microcomputer and spreadsheet program was employed to develop useful information quickly.

The method was straightforward. A spreadsheet template was prepared containing the MLRA numbers in the first column. It could then be used to enter such data as total range acres, acres needing various types of conservation treatment, or any other factor in adjoining columns. The spreadsheet was programmed to calculate the national totals by adding each column. If that matched the total provided by the SCS summary, the entries were assumed correct; if not, errors were located and corrected.

With the raw numbers entered, other columns on the spreadsheet could be programmed to calculate percentages or any other necessary calculation. The particular spreadsheet program used (SuperCalc2) was also capable of rearranging the entries in ascending or descending order very rapidly. In this way, the MLRAs could be ranked according to any acreage or percentage characteristics. Through the use of this increasingly common tool, a number of comparative tables were generated in a very short time (<u>see</u> Table 15), and MLRAs could be chosen for

TABLE 15 Rangeland Characteristics as Shown in the 1982 NRI, Ranked by Total Acres of Range in MLRA (1,000 Acres)

Rank	MLRA	Total Acres	Adequately Protected Acres	Adequately Protected Percent	Needs Erosion Control Acres	Needs Erosion Control Percent	Needs Re-Establishment Acres	Needs Re-Establishment Percent
1	42	24,374.9	7,412.0	30.41	2,110.6	8.66	268.9	1.10
2	78	21,303.9	6,986.3	32.79	765.3	3.59	408.2	1.92
3	81	21,244.8	3,341.4	15.73	239.0	1.12	437.3	2.06
4	58A	16,794.2	7,169.2	42.69	220.9	1.32	123.5	0.74
5	70	15,307.4	4,715.9	30.81	2,477.1	16.18	162.2	1.06
6	77	13,876.1	4,977.0	35.87	1,945.8	14.02	288.5	2.08
7	35	13,672.0	3,010.6	22.02	832.8	6.09	1,020.6	7.46
8	65	11,389.3	8,416.8	73.90	33.3	0.29	195.7	1.72
9	67	9,963.2	3,384.5	33.97	1,261.8	12.66	196.9	1.98
10	34	9,773.0	2,368.6	24.24	381.9	3.91	91.7	0.94

SOURCE: Derived from 1982 NRI.

study on the basis of known characteristics and rankings for selected criteria.

Analyses of this kind can use the data from the SCS summaries on an ordinary office microcomputer to develop information that is likely to be adequate for a wide variety of policy and program planning needs, as well as giving useful insights that can be helpful in public information programs. While this will not always satisfy the precision requirements of researchers, it is both inexpensive and efficient, and should not be overlooked as an opportunity that is available, for the first time, with the 1982 NRI.

Discussion

Kenneth G. Renard

This is a most thorough and thought-provoking summary and analysis of the 1982 National Resources Inventory (NRI) survey for erosion from range and forestland. The assessments contained in Sampson's paper suggest additional analyses and summaries that would be worthwhile and supportive of Soil Conservation Service (SCS) targeting efforts.

The Society for Range Management (SRM) has gone on record as being opposed to use of the Universal Soil Loss Equation (USLE). The group (Schuster, 1984) contends that "until technology is developed to replace it...the USLE is inapplicable for assessing the resources on rangelands." The SRM further "encourages the U.S. Department of Agriculture (USDA) to adopt proven and acceptable techniques for evaluating vegetation as a more accurate and earlier indication of degradation of the total rangeland resource." Such statements have done a real disservice to the USLE, which was never intended to assess anything but the erosion that would be expected over a relatively long time.

Use of the USLE in the 1982 NRI evaluation and its analysis by papers such as this one are valid applications of the USLE for targeting the use of resources--dollars and personnel. The question that remains is whether the USLE has received sufficient verification and validation for use on rangeland.

Some of the earliest measurements of soil erosion were made by A. W. Sampson (Sampson and Weyl, 1918), assisted by L. H. Weyl, E. V. Storm, and C. L. Forsling, on overgrazed rangelands in central Utah. These studies, and research by Chapline (1929), illustrated how overgrazing allowed erosion to reduce soil fertility and waterholding capacity. Unfortunately, erosion research on rangeland languished from the time of these early efforts until the 1970s. Concern for the ecological health of rangeland grew with general concern for the environment that developed during the late 1960s and 1970s. Excessive erosion was again recognized as being detrimental to rangelands as well as other agricultural lands. Consequently, current management plans for rangelands frequently contain analyses on how management alternatives would affect erosion. Since research has provided little information on erosion associated with rangeland activities, technology from other geographic areas was adapted to estimate erosion on rangeland. In particular, the USLE, which has been used successfully on cropland since the early 1960s, was adopted to estimate erosion on rangeland.

Had computer technology been available in the 1940s, current erosion prediction methods might look more like the theory contained in Ellison's classic paper (1947) than like the empirical form of the USLE. The USLE and its predecessors were very much structured to be "user friendly," because by the early 1950s erosion equations were accepted by the USDA-SCS as a tool for tailoring erosion control practices to the needs of specific fields and farms. Unfortunately, during this period no comparable erosion research program on rangelands in the western United States was conducted, and thus recent efforts to develop erosion methods for rangelands have not had an extensive data base on which to draw.

Although the USLE was being applied on a limited basis prior to its 1965 release in Agriculture Handbook 282 (Wischmeier and Smith, 1965), the SCS and other agencies soon switched from the regional agronomic planning concepts for erosion abatement to the USLE, and by the mid-1970s there was an interest in using the technology on western rangelands. Thus, requests were made for a "best estimate" approach for the cover-management factor [Wischmeier then developed Table 10 in Agriculture Handbook 537 (Wischmeier and Smith, 1978)] until such time that research could provide data for a similar table or an alternative.

Table 1 presents a list of some of the material that has appeared in the scientific literature over the past few years regarding application of the USLE to range

TABLE 1 Research Evaluating USLE Performance on Rangelands

Reference	Area of Work	Comments
Renard et al., 1974	Arizona	Used small watersheds; significant channel erosion
Renard and Simanton, 1975	Arizona, New Mexico	Explored estimation of erosion factor
Osborn et al., 1976	Arizona, New Mexico	Showed importance of stone surface cover
Simanton et al., 1977	Arizona	Showed effect of root plowing and reseeding on erosion control
Verma et al., 1977	Arizona	Measured erosion from disturbed and natural plots with artificial or simulated rainfall
Johnson et al., 1980	Idaho	Used canopy and ground cover to compute potential erosion for sagebrush control
Renard, 1980	Arizona	Compared numerous sediment yield formulae
Simanton et al., 1980	Arizona	Applied to small watersheds on storm basis
Trieste and Gifford, 1980	Utah	Used small plots with rainfall simulator; suggested USLE did not apply well to rangelands
Foster et al., 1981	Utah	Discussed applicability of USLE to rangelands
Dissmeyer, 1982	New Mexico	Used subfactor approach in evaluating C (cover) on rangeland
Hart, 1982	Utah	Measured erosion on sagebrush plots with a rainfall simulator

TABLE 1 (Continued)

Reference	Area of Work	Comments
McCool, 1982	Washington	Theoretical analysis of slope length-steepness factor
Renard and Stone, 1982	Arizona	Correlation of USLE estimates with stock pond yields
Simanton and Renard, 1982	Arizona, New Mexico	Evaluated erosivity of air-mass thunderstorms
Williams, 1982	Texas, Oklahoma, Iowa, New Mexico	Estimated sediment yield from mixed cover watersheds with modified USLE
Trott and Singer, 1983	California	USLE soil erodibility factor should consider soil mineralogy
Hart, 1984	Utah	Fair agreement of USLE with simulated rainfall data; slope factor needs adjustment
Simanton et al., 1984	Arizona	Measured erosion reduction caused by stone surface cover
Smith et al., 1984	Texas, Oklahoma	Sediment yield estimates with modified USLE, on watersheds less than 122 hectares and on watersheds with mixed land uses
Tracy et al., 1984	Arizona	Measured drop-size distribution of air-mass thunderstorms for use in evaluating erosivity
Johnson et al., 1984	Idaho, Nevada	Used rainfall simulator and found interpretation of C on ungrazed areas needed refinement

conditions. Despite this considerable attention, many problems remain unresolved, although analysts are getting much closer to being comfortable with this technology.

Two years ago, the Bureau of Land Management asked a number of Agricultural Research Service (ARS) and university researchers to develop a handbook for the application of the USLE on rangelands. The response to this request made several issues apparent: (1) A major effort was needed to improve the evaluation of C, the cover-management factor; (2) this improvement could best be accomplished by using a subfactor approach for evaluating C; (3) there are problems in assessing the orographic effects of precipitation in the form of rain and snow on EI, the storm kinetic energy times maximum 30-minute intensity; (4) snowmelt is a problem in estimating erosion; (5) frozen soils and the freeze/thaw cycles that occur frequently on rangeland represent a special problem; and (6) slope length and steepness are often greater than that encountered on cultivated cropland.

Time does not permit treatment of all of these problems. Rather, a discussion of the subfactor approach for evaluation of C will be presented. The procedure is very similar to that presented by Dissmeyer (1982) and Dissmeyer and Foster (1981) for forestland in the southeastern United States and now used elsewhere. The cover-management factor for rangeland is given as (J. M. Laflen, USDA-ARS, Ames, Iowa, personal communication, 1984):

$$C = (PLU)(PC)(SC)(SR), \tag{1}$$

where PLU is a prior land use subfactor, PC is a plant canopy subfactor, SC is a surface cover subfactor, and SR is a surface roughness subfactor. The individual subfactors can be obtained as follows:

$$PLU = 0.45 \, EXP(-.012 \, RS), \tag{2}$$

where RS is the mass of roots and residue (kilograms/hectare/millimeter of depth) in the surface 100 millimeters of soil. At present, there are no adjustments in this subfactor to account for differences in grazing intensity. However, the coefficient 0.45 does express the long-term consolidation effects occurring on rangeland due to grazing. Other grazing effects, such as reduced canopy cover, different surface cover, or roughness changes, are reflected in other subfactors.

If the rangeland is tilled, the PLU is assumed to follow this relationship for 7 years:

$$PLU = (1 - 0.08\,Y)\,EXP(-.012\,RS). \qquad (3)$$

The relationship of plant canopy to soil erosion was taken from Wischmeier and Smith (1978) and given as:

$$PC = 1 - FC(EXP - 0.34\,H), \qquad (4)$$

where FC is the fraction of the land surface covered by canopy and H is the average canopy height (meters).

Surface cover creates small dams where runoff is temporarily ponded and eroded sediment may be deposited. The surface cover factor is expressed as:

$$SC = EXP(-3.5\,M), \qquad (5)$$

where M is the fraction of the land surface covered by nonerodible material such as litter, rock, and growing vegetation.

Surface roughness influences soil erosion by reducing runoff volume and velocity, and by ponding surface runoff to cause deposition. The roughness of a surface is expressed as the standard deviation among heights along the soil surface perpendicular to the slope. The algorithm used to compute the subfactor is:

$$SR = EXP[-.026(RB - 6)(1 - EXP(-.035\,RS))], \qquad (6)$$

where RB is surface roughness and RS is as defined earlier. Tables and pictures for estimating RB are given in the document to assist the user in selecting the appropriate value for the condition being considered.

Of concern is how much changes in the USLE parameters resulting from new USLE information might affect the rangeland summaries in the NRI. Because the research is unlikely to be completed until after the targeting objectives of the Resource Conservation Act process are in effect using the 1982 NRI data, perhaps the answer will never be known. It does seem likely that confidence in the numbers obtained would be improved and that professional societies like the SRM will be more amenable to working on the rangeland resource problem. Likewise, the USLE technology used for the 1982 NRI is based on fundamental concepts and, as such, should provide reasonable planning/inventory data on water erosion.

SOIL LOSS TOLERANCES

Sampson's paper refers to 83 percent of the rangeland with estimated soil loss at or below the soil loss tolerance limit (T value). Furthermore, 6.4 percent of the area had loss rates between T and 2T, and 10.8 percent had soil erosion in excess of 2T. Presumably a T value for rangeland of 5 tons/acre/year was used. Is there sufficient information on soil formation processes on rangeland to establish T values? Soil morphologists have noted that many of the soils on Walnut Gulch in southeastern Arizona are not soil (based on their experience in more humid areas) but rather partly weathered parent geologic material. Thus, given the dry conditions, low organic matter, and other factors, soil loss may not be affordable (in a noneconomic sense). But geologic erosion has always been taking place in such areas. The question, then, is how significant current erosion is relative to geologic erosion. In fact, in many rangeland areas, erosion from the rill/interrill areas is not the major sediment source. It is the material coming from headcuts, arroyo entrenching, channel degrading and widening, or other sources that is the major contributor to the downstream sediment yield. Thus, unless land management alters the runoff distribution, downstream sedimentation may not be rectified. And this is not even a part of the assessment of the NRI.

The wind erosion estimates in the section on rangeland of this paper are very interesting. Like the USLE, the Wind Erosion Equation has certainly had minimal testing on western rangelands, with the exception of some work in Texas and New Mexico. The wind erosion problem specifically cited in Table 1 for MLRA 30 is an interesting one. The area, on both sides of the lower Colorado River near the U.S.-Mexico border, is quite arid and contains many sand dunes and an extreme shortage of vegetation in the nonirrigated condition. Further, desert pavements are quite common in the area. Were allowances for the gravel on the soil surface made? The I value (soil erodibility) selected was probably based on soil texture determined without considering the gravel. Dr. Leon Lyles, director of the ARS Wind Erosion Laboratory at Manhattan, Kansas, has stated that the gravel should be considered in the textural evaluation and thus in the I value that might be selected (personal communication, 1984). Adequate protection with vegetation treatments is extremely difficult in such areas when moisture is so limiting.

REFERENCES

Chapline, W. R. 1929. Erosion on rangeland. J. Am. Soc. Agron. 21:423-429.

Dissmeyer, G. E. 1982. Report on application of the USLE to desert brush and rangeland conditions. Pp. 214-225 in Proc. of Workshop on Estimating Erosion and Sediment Yields on Rangelands. ARM-W-26. Washington, D.C.: USDA Agricultural Research Service.

Dissmeyer, G. E., and G. R. Foster. 1981. Estimating the cover-management factor (C) in the Universal Soil Loss Equation for forest conditions. J. Soil Water Conserv. 36:235-240.

Ellison, W. D. 1947. Soil erosion studies. Agric. Engineer. 28:145-146, 197-201, 245-248, 297-300, 349-351, 402-405, 442-444.

Foster, G. R., J. R. Simanton, K. G. Renard, L. J. Lane, and H. B. Osborn. 1981. Viewpoint: Application of the USLE to rangelands on a per-storm basis. J. Range Manage. 34:161-166.

Hart, G. E. 1982. A test of the USLE on bare and sagebrush plots in Utah. Pp. 101-105 in Proc. of Workshop on Estimating Erosion and Sediment Yield on Rangelands. ARM-W-26. Washington, D.C.: USDA Agricultural Research Service.

Hart, G. E. 1984. Erosion from simulated rainfall on mountain rangeland in Utah. J. Soil Water Conserv. 39:330-334.

McCool, D. K. 1982. Effects of slope length and steepness on soil erosion from rangelands. Pp. 73-95 in Proc. of Workshop on Estimating Erosion and Sediment Yield on Rangelands. ARM-W-26. Washington, D.C.: USDA Agricultural Research Service.

Johnson, C. W., G. A. Schumaker, and J. P. Smith. 1980. Effects of grazing and sagebrush control on potential erosion. J. Range Manage. 33(6):451-454.

Johnson, C. W., M. R. Savabi, and S. A. Loomis. 1984. Rangeland erosion measurements for the USLE. Trans. ASAE 127:1313-1320.

Osborn, H. B., J. R. Simanton, and K. G. Renard. 1976. Use of the Universal Soil Loss Equation in the semiarid Southwest. Pp. 41-49 in Soil Erosion: Prediction and Control. Special Pub. 21. Ankeny, Iowa: Soil Conservation Society of America.

Renard, K. G. 1980. Estimating erosion and sediment yields from rangelands. Pp. 164-175 in Proc. of Am. Soc. Civil Engineers Symp. on Watershed Management. Boise, Idaho: American Society of Civil Engineers.

Renard, K. G., and J. R. Simanton. 1975. Thunderstorm precipitation effects on the rainfall-erosion index of the Universal Soil Loss Equation. Hydrol. Water Res. Ariz. Southwest 5:47-55.

Renard, K. G., and J. J. Stone. 1982. Sediment yield from small arid rangeland watersheds. Pp. 129-144 in Proc. of Workshop on Estimating Erosion and Sediment Yield on Rangelands. ARM-W-26. Washington, D.C.: USDA Agricultural Research Service.

Renard, K. G., J. R. Simanton, and H. B. Osborn. 1974. Applicability of the Universal Soil Loss Equation to semiarid rangeland conditions in the Southwest. Hydrol. Water Res. Ariz. Southwest 4:18-32.

Sampson, A. W., and L. H. Weyl. 1918. Range Preservation and Its Relation to Erosion Control on Western Grazing Lands. U.S. Department of Agriculture Bulletin No. 675. Washington, D.C.: U.S. Government Printing Office.

Schuster, J. L. 1984. Position statements by SRM. Rangelands 6(3):139-140.

Simanton, J. R., and K. G. Renard. 1982. The USLE rainfall factor for southwestern U.S. rangelands. Pp. 50-62 in Proc. of Workshop on Estimating Erosion and Sediment Yield on Rangelands. ARM-W-26. Washington, D.C.: USDA Agricultural Research Service.

Simanton, J. R., H. B. Osborn, and K. G. Renard. 1977. Effects of brush to grass conversion on the hydrology and erosion of a semiarid southwestern rangeland watershed. Hydrol. Water Res. Ariz. Southwest 7:249-256.

Simanton, J. R., H. B. Osborn, and K. G. Renard. 1980. Application of the USLE to southwestern rangelands. Hydrol. Water Res. Ariz. Southwest 10:213-220.

Simanton, J. R., E. Rawitz, and E. D. Shirley. 1984. The effects of rock fragments on erosion of semiarid rangeland soils. Pp. 65-72 in Erosion and Productivity of Soils Containing Rock Fragments. Special Pub. 13. Madison, Wis.: Soil Science Society of America.

Smith, S. J., J. R. Williams, R. G. Menzel, and G. A. Coleman. 1984. Prediction of sediment yield from southern plains grasslands with the modified Universal Soil Loss Equation. J. Range Manage. 37(4):295-297.

Tracy, F. C., K. G. Renard, and M. M. Fogel. 1984. Rainfall energy characteristics for southeastern Arizona. Pp. 559-566 in Water--Today and Tomorrow. Proc. of Am. Soc. Civil Engineers Irrig. and Drain. Div. Specialty Conference July 24-26, Flagstaff, Arizona.

Trieste, D. J., and G. F. Gifford. 1980. Application of the Universal Soil Loss Equation to rangelands on a per-storm basis. J. Range Manage. 33(1):66-70.

Trott, K. G., and M. J. Singer. 1983. Relative erodibility of 20 California range and forest soils. Soil Sci. Soc. Am. J. 47:753-759.

Verma, T. K., J. L. Thames, and J. E. Mills. 1977. Soil erosion and sediment control on the reclaimed coal mine lands of the semiarid Southwest. Hydrol. Water Res. Ariz. Southwest 7:41-47.

Williams, J. R. 1982. Testing the modified Universal Soil Loss Equation. Pp. 157-165 in Proc. of Workshop on Estimating Erosion and Sediment Yield on Rangelands. ARM-W-26. Washington, D.C.: USDA Agricultural Research Service.

Wischmeier, W. H., and D. D. Smith. 1965. Predicting Rainfall-Erosion Losses from Cropland East of the Rocky Mountains--Guide for Selection of Practices for Soil and Water Conservation. Agriculture Handbook No. 282. Washington, D.C.: U.S. Government Printing Office.

Wischmeier, W. H., and D. D. Smith. 1978. Predicting Rainfall-Erosion Losses--A Guide to Conservation Planning. Agriculture Handbook No. 537. Washington, D.C.: U.S. Government Printing Office.

7

Erosion Control Practices: The Impact of Actual Versus Most Effective Use

Paul E. Rosenberry and *Burton C. English*

For decades, the United States has both enjoyed and suffered from the ability of its agricultural sector to produce more than was demanded domestically or than could be sold abroad at a profit. Consumers benefited from this insofar as food and fiber prices were below those that would otherwise have prevailed. The agricultural sector, on the other hand, suffered from low prices relative to production costs and the resulting downward pressure on net farm income. Most farm programs have sought to control acreages in crop production, thus reducing the cropland base. Diminishing profits encouraged farmers to intensify planting on acres still in production and to convert noncropland to cropland in order to qualify more acres for production control programs. Lower or nonexistent profits on noncropland uses further accelerated cropland conversion. The farmer's ability to increase yields of controlled crops has tended to offset the programs that aimed to reduce agricultural production.

Individual farmers have responded well to the signals they have received. The government has encouraged maximum production on every acre that can qualify. As a result, farmers have been trying to produce their way out of a surplus situation. Acreage control programs have thus resulted in:

- intensive cropland use;
- increased conversion of noncropland to cropland;
- increased misuse of natural resources;
- increased capitalization of land and farm equipment;

- decreasing net incomes as prices of farm products have failed, even with price support programs, to keep pace with production costs;
- the expense of administering acreage control programs; and
- the expense of storing surplus commodities.

Since the early 1970s, increased foreign demand reduced agricultural product stockpiles and signaled farmers, through higher prices, to increase production. Financiers encouraged farmers to borrow capital. However, the ability of foreign agricultural producers to sell commodities below the U.S. support price forced American exporters to be residual suppliers. Domestically, tightening supply-demand markets for capital (largely a result of increasing national debt and unfavorable balance of payments) caused interest rates to rise dramatically.

The prospects for the rest of the 1980s suggest a continuing variable relationship between the demand for and the supply of agricultural commodities. World population continues to increase and inherent soil productivity continues to decline. Uncontrollable factors such as weather or foreign political instability are likely to keep foreign demand for U.S. farm commodities unpredictable. A clear understanding of how soils are being protected or depleted is therefore essential.

This paper explores the extent and severity of sheet and rill erosion, the extent to which conservation practices have reduced potential erosion, and the ability of conservation practices to resolve sheet and rill erosion problems.

The national survey of lands used primarily for agricultural purposes by the Soil Conservation Service (SCS) is designed to yield statistically reliable estimates at the national, state, and Major Land Resource Area (MLRA) levels. Important parts of the National Resources Inventory (NRI) are the observations for each element in the Universal Soil Loss Equation (USLE) and the resulting estimates of the rate of sheet and rill erosion at each sample point in the survey. (For discussions of the elements of the equation--RKLSCP--*see* Heimlich and Bills, Pierce et al., and Runge et al., this volume.)

$$A = RKLSCP$$

TABLE 1 Cropland Uses by Sheet and Rill Erosion Rate, United States, 1982 (1,000 Acres)

USLE (Tons/ Acre/Year)	Row Crops	Close-Grown Crops	Hay	Other Crops	Total[a]
0.0--<5.0	142,963.1	97,560.6	36,548.7	50,233.5	327,306.2
5.0--<10.0	34,227.3	10,980.8	642.0	4,737.0	50,587.3
10.0--<15.0	11,048.3	3,403.5	180.1	1,608.0	16,240.2
15.0--<20.0	5,799.5	1,541.2	66.6	599.0	8,006.5
20.0--<25.0	3,409.7	722.1	25.2	339.2	4,496.5
25.0--<30.0	2,378.0	440.5	23.3	204.3	3,046.5
30.0--<35.0	1,527.7	296.8	7.5	123.1	1,955.3
35.0--<40.0	1,162.7	215.0	0.5	81.0	1,459.5
40.0--<50.0	1,470.8	177.7	5.1	87.1	1,741.0
50.0--<75.0	1,517.0	207.2	3.5	76.6	1,804.5
75.0--<100.0	460.3	39.3	1.0	28.2	529.1
100.0 & up	319.5	29.5	0.0	22.6	371.8
Total	206,285.0	115,615.2	37,504.2	58,140.6	417,545.3
Erosion (million tons)	1,272.4	372.3	25.8	160.7	1,831.2
Average erosion (tons/acre/year)	6.1	3.2	0.6	2.7	4.3

[a]Figures have been rounded off and do not total to an exact number.

SOURCE: 1982 NRI.

Generally, except for terracing, the first four factors in the equation reflect the natural potential of land to erode, while the last two factors reflect managerial decisions that determine how much potential erosion is actually realized. The exception is that terraces shorten slope lengths and, in certain cases, lower overall slope. Nevertheless, the assumption that RKLS represents the natural erosion potential of a soil is valid because terraces only occurred on 7 percent of the sample points in the 1982 NRI.

Table 1 summarizes the sheet and rill erosion on cropland documented in the 1982 NRI for the United States. These rates are annual averages of soil movement within a field or sample point. It should be noted that these data may differ from other published NRI results in that only privately owned land is used. Also, pastureland and

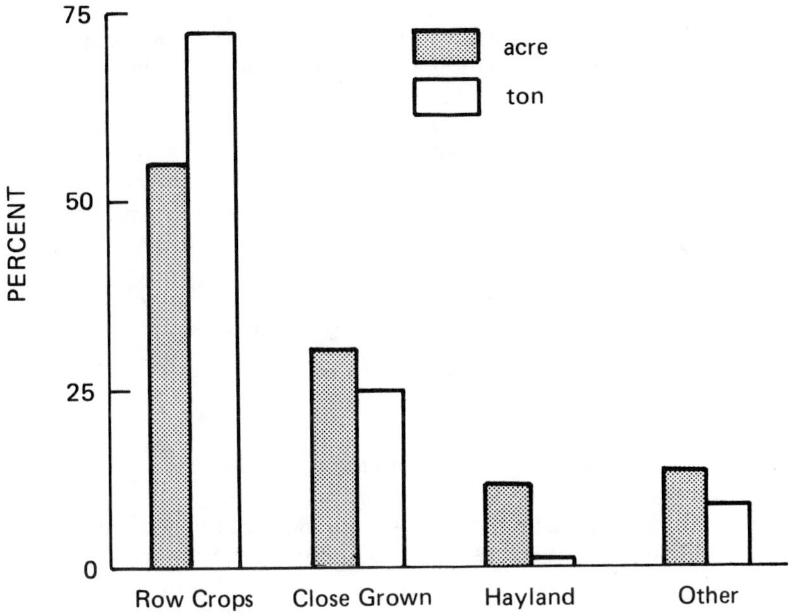

FIGURE 1 Cropland use and erosion in the United States in 1982.

native pastureland with tillable cropland history are included in the other crops category.

The NRI data indicate that 417.5 million acres are readily available for crop production in the United States. Soil movement from sheet and rill erosion is 4.3 tons/acre/year overall. Even land used for row crops has an erosion rate of only 6.1 tons/acre/year. Row crops are planted on almost half the cropland but account for almost 70 percent of sheet and rill erosion (see Figure 1). Although row cropland comprises almost 44 percent of U.S. soil loss at fewer than 5 tons/acre/year, it accounts for almost 87 percent of the loss at 75 to less than 100 tons/acre/year. The other cropland categories all have erosion rates below the average. Over 78 percent of U.S. cropland registers sheet and rill erosion of less than 5 tons/acre/year.

For the whole country, land eroding at less than 5 tons/acre/year accounts for about one-third the soil loss in the 48 states. Lands eroding at the rate of 5 to 15 tons/acre/year account for another one-third. The remaining one-third of the lands have annual sheet and rill erosion rates greater than 15 tons/acre. They have one-third of total erosion, but represent only 6 percent of total cropland. If these 25 million acres could be taken out of cultivation, excess erosion could be decreased by almost one-third. It should be noted that these lands may be intermingled with other croplands, and they may be only small parts of fields. Such small tracts may seem insignificant to individual managers, and not farming them may be more costly in time or money. These lands would seem to be a prime target in a program to reduce U.S. erosion.

FACTORS IN CROPLAND EROSION

As Table 1 suggests, averages do not tell the whole story. Valuable insights can be gained, however, by reviewing the factors responsible for sheet and rill erosion on land used to grow row crops and small grains in 1982.

The Impact of C and P Factors

As indicated, the rainfall, erodibility, slope length, and slope gradient elements of the USLE can be considered as naturally occurring factors in the sheet and rill erosion process. The equation is designed so that the product of R, K, L, and S can be used, under certain assumptions, as an estimate of what sheet and rill erosion would be in the absence of the cover and management factor (C), the supporting practice factor (P), and the impact of the shorter slope length of terraces. The RKLS product serves as an estimate of erosion in this case for two reasons: First, land is assumed to be tilled fallow with no plant or residue production. Second, tillage is assumed to occur up and down the field slopes.

Tables 2 and 3 show cropland acreage and tons of erosion by RKLS grouping, indicating the distribution of natural potential for sheet and rill erosion as revealed by the NRI sample points. The average RKLS for all cropland is calculated to be about 21.8 tons/acre/year when the sample observations are weighted by the acreage

TABLE 2 Cropland Uses by RKLS Factor, United States, 1982 (Million Acres)

RKLS (Tons/ Acre/Year)	Row Crops	Close-Grown Crops	Hay	Other Crops	Total
0.0--<5.0	44.8	38.0	12.0	19.4	114.3
5.0--<10.0	56.7	28.0	6.5	13.5	104.8
10.0--<15.0	31.6	16.6	3.0	6.7	58.0
15.0--<20.0	18.1	9.3	2.6	4.1	34.1
20.0--<25.0	10.8	6.1	1.5	2.4	20.8
25.0--<30.0	7.3	4.2	1.4	1.8	14.7
30.0--<35.0	5.4	2.9	1.0	1.3	10.5
35.0--<40.0	4.0	1.9	1.0	1.0	7.8
40.0--<50.0	6.0	2.6	1.4	1.5	11.5
50.0--<75.0	8.9	3.2	2.1	2.3	16.6
75.0--<100.0	4.7	1.4	1.4	1.3	8.8
100.0 & up	7.9	1.6	3.4	2.7	15.6
Total	206.2	115.8	37.3	58.0	417.5

SOURCE: 1982 NRI.

they represent, compared with the actual rate of 4.3 tons/acre/year (<u>see</u> Table 1). This suggests that if the 417.5 million acres used for cropland had been in fallow and had been tilled up and down the slopes, an average 21.8 tons/acre/year would be lost to sheet and rill erosion. The difference is due to the impact of current managerial decisions regarding cover and management practices and to shortened slope lengths of terraces.

A comparison of the average tons per acre in Table 4 shows that row crops are being grown on nearly average cropland, a relationship explained in part by the fact that row crops comprise about 50 percent of cropland. Close-grown crops are planted on the least erodible land overall, while hay is grown on the most erodible (<u>see</u> Table 4). The removal of C and P factors changes total tons of erosion on hayland from 25.8 million (<u>see</u> Table 1) to 1.3 billion tons (<u>see</u> Table 3). Land in row crops would register the largest change in tonnage, with an increase from 1.2 billion (<u>see</u> Table 1) to 4.6 billion tons (<u>see</u> Table 3).

TABLE 3 Sheet and Rill Erosion if Cropland Were in Tilled Fallow, by RKLS Factor, United States, 1982 (Million Tons)

RKLS (Tons/Acre/Year)	Row Crops	Close-Grown Crops	Hay	Other Crops	Total[a]
0.0--<5	140	108	26	52	326
5.0--<10.0	415	204	48	99	766
10.0--<15.0	389	204	38	83	713
15.0--<20.0	313	161	47	71	592
20.0--<25.0	241	137	35	54	466
25.0--<30.0	200	114	40	51	404
30.0--<35.0	175	93	32	42	342
35.0--<40.0	150	70	36	38	294
40.0--<50.0	542	194	131	142	1,008
50.0--<75.0					
75.0--<100	402	118	129	109	758
100.0 & up	1,363	266	698	621	2,949
Total	4,600	1,784	1,323	1,428	9,135
Average RKLS	22.3	15.4	35.2	24.5	21.8

[a]Totals are rounded off.

SOURCE: 1982 NRI.

Table 5 shows the distribution of cover and management conditions (the C factor) on land cultivated in 1982. The weighted average value of C reported in the NRI for cropland in 1982 was .26 (*see* Table 6). Thus, overall plant matter and residues reduced annual average sheet and rill erosion from the 21.8 tons/acre that would have prevailed under conditions of tilled fallow (*see* Table 2) to around 6 tons/acre.

The C-factor value for all cropland could be expected to mirror that of row crops, given the dominance of the latter. However, a distinct difference can be seen when C-factor values greater than .45 tend to be row crops, and from there on the distributions are similar (*see* Figure 2).

A comparison of C-factor values indicates that plant matter and residue are most effective in reducing erosion rates on hayland and least effective on land in row crops

TABLE 4 Impact of C and P Factors in USLE Estimates for Sheet and Rill Erosion on Cropland, United States, 1982

Crops	Inherent Erosion Potential (RKLS) (Tons/Acre/Year)	Actual Erosion (RKLSCP) (Tons/Acre/Year)	Reduction Factor	Percent Reduction
Row crops	22.3	6.1	3.7	72.6
Close-grown crops	15.4	3.2	4.8	79.2
Hay	35.2	0.6	58.7	98.2
Other	24.5	2.7	9.1	88.9
All cropland	21.8	4.3	5.1	80.2

SOURCE: 1982 NRI.

(see Tables 4 and 6). This is not surprising, but the overall magnitude of the effectiveness of plant and residue to lower erosion rates should be noted. For example, the overall impact of hay is to lower RKLS values of 125 tons per acre to 5 tons per acre.

Another way to lower soil erosion is through supporting practices that lower the value of the P factor. Table 7 shows the distribution of major land uses by supporting practice groups on land cultivated in 1982. The P-factor values reflect the extent to which erosion is further reduced beyond that brought about by plant and residue conditions (the C factor). The weighted average value of P in the 1982 NRI for all cropland was .91 (see Table 6). Thus, the overall impact of supporting practices was to reduce sheet and rill erosion by 9 percent (1.0 - .91)--about 2 tons/acre. One reason for this small impact is that only 40 percent of cropland has a supporting practice that lowers P-factor values.

A comparison of P-factor values across all land uses indicates that supporting practices do not vary by land uses. There are a few acres reported in the .45 to less than .70 range, reflecting contouring and stripcropping, but the dominant range is the .90 to 1.0 category (see Table 7).

A comparison of weighted average C and P factors (see Figure 3) illustrates the overall impact of plant residue

TABLE 5 Acreage and Distribution of Cropland Uses by C Factor, United States, 1982

C Factor	Row Crops Acres (Million)	Row Crops Percent	Close-Grown Crops Acres (Million)	Close-Grown Crops Percent	Hay Acres (Million)	Hay Percent	Other Crops Acres (Million)	Other Crops Percent	Total Acres (Million)	Total Percent
.00--<.02	0.6	0.3	0.8	0.7	25.5	68.0	8.0	13.8	34.9	8.4
.02--<.05	2.0	1.0	2.3	2.0	5.6	15.0	4.5	7.7	14.4	3.4
.05--<.07	1.7	0.8	1.7	1.4	1.4	3.8	1.8	3.1	6.6	1.6
.07--<.10	4.8	2.3	6.9	6.0	2.4	6.3	5.6	9.1	19.7	4.7
.10--<.15	12.3	6.0	17.8	15.4	2.6	6.9	7.9	13.5	40.5	9.7
.15--<.20	16.9	8.2	20.6	17.8	0.0	0.0	7.4	12.8	45.0	10.8
.20--<.25	18.3	8.9	20.6	17.8	0.0	0.0	5.7	9.9	44.7	10.7
.25--<.30	21.6	10.5	16.1	13.9	0.0	0.0	4.3	7.4	42.0	10.1
.30--<.35	30.5	14.7	12.8	11.1	0.0	0.0	3.7	6.4	47.0	11.2
.35--<.40	39.4	19.1	7.6	6.6	0.0	0.0	3.5	6.1	50.6	12.1
.40--<.45	25.2	12.2	4.0	3.4	0.0	0.0	2.4	4.2	31.5	7.6
.45--<.50	18.1	8.8	2.2	1.9	0.0	0.0	1.5	2.5	21.8	5.2
.50--<.60	9.5	4.6	1.9	1.7	0.0	0.0	1.1	1.9	12.5	3.0
.60--<.70	5.2	2.5	0.3	0.3	0.0	0.0	0.5	0.9	6.0	1.4
.70--<.80	0.2	0.1	0.0	0.0	0.0	0.0	0.9	0.1	0.2	0.0
.80--<.90	0.0	0.0	0.0	0.0	0.0	0.0	0.1	0.1	0.1	0.0
.90--1.00	0.0	0.0	0.0	0.0	0.0	0.0	0.0	0.1	0.0	0.0
Total[a]	206.3	100.0	115.6	100.0	37.5	100.0	58.1	100.0	417.5	100.00

[a]Figures have been rounded off and do not total to an exact number.

SOURCE: 1982 NRI.

TABLE 6 C and P Factor Weighted Averages by Land Use, United States, 1982

Land Use	C Factor	P Factor
Row crops	.28	.86
Close-grown crops	.20	.87
Hay	.04	.81
Other crops	.19	.88
Total	.26	.91

SOURCE: 1982 NRI.

management over supporting practice on U.S. totals. The white bars representing supporting practices are all close to 1.0, thus having minimal impact on national totals. The black bars are much closer to zero, indicating greater ability to decrease erosion. Thus, from a national policy viewpoint, it would seem that land use changes reflecting plant and residue changes are more important than supporting practices. This point has received little attention because comparisons are often made between supporting practices given plant and residue cover.

The Impact of Minimum Tillage, Contour Farming, Stripcropping, and Terracing

There are about 167 million acres with some form of minimum tillage, contour, stripcropping, and terrace systems (<u>see</u> Table 8). This amounts to 81 percent of row cropland and 40 percent of all cropland. Minimum tillage dominates conservation practice use and is concentrated on row-crop acres. Contour and terrace systems rank next, with about one-third as much acreage as minimum tillage.

Some of the varying effectiveness of minimum tillage is shown in Table 9. The percentage distribution shows the value of the C factor across the complete range of possibilities, with a modal mean at the .15 to less than .20 category (<u>see</u> Figure 4). The acreage with the lowest

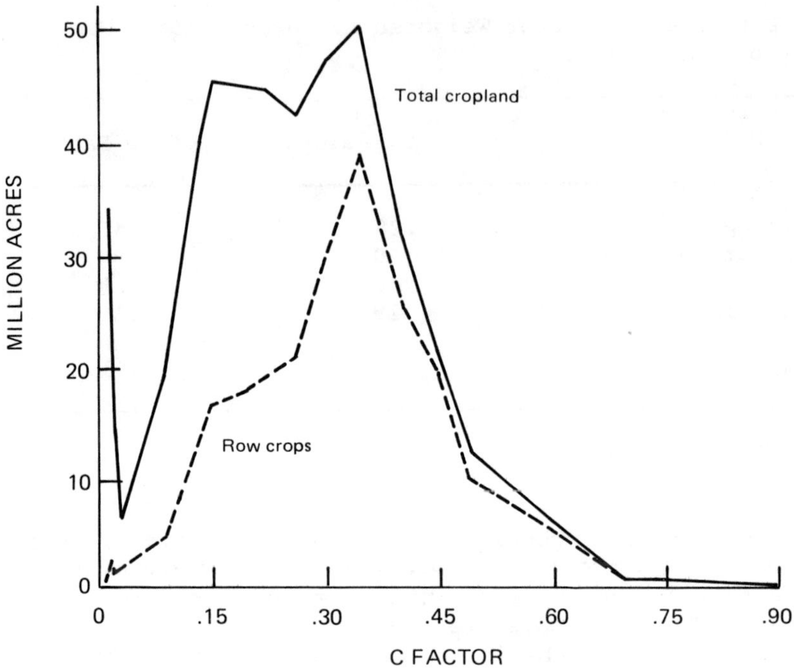

FIGURE 2 Distribution of row crops and total cropland by C factor groups in the United States, 1982.

SOURCE: 1982 NRI.

C factors reflects the cropland with high concentration of residue or sod-planted crops. The other extreme is not so easily explained. It would appear the residue and root structure of the vegetation present are not significantly different from tilled fallow. The main point is still clear: Namely, that the value of the C factor for minimum tillage systems can vary from extremely effective to only as effective as fall-plowed row crops. This variation is probably caused by the wide range of residue quantity that can be left on the soil surface and still qualify as minimum tillage. The current SCS definition of conservation tillage is to have 30 to 80 percent of the ground covered with residue after planting. This definition may not have been used in the survey. Also, it is not possible for all survey interviews to be conducted at planting time.

TABLE 7 Acreage and Distribution of Cropland Uses by P Factor, United States, 1982

P Factor	Row Crops Acres (Million)	Row Crops Percent	Close-Grown Crops Acres (Million)	Close-Grown Crops Percent	Hay Acres (Million)	Hay Percent	Other Crops Acres (Million)	Other Crops Percent	Total Acres (Million)	Total Percent
.00--<.02	0.0	0.0	0.0	0.0	0.0	0.0	0.0	0.0	0.0	0.0
.02--<.05	0.0	0.0	0.0	0.0	0.0	0.0	0.0	0.0	0.0	0.0
.05--<.07	0.0	0.0	0.0	0.0	0.0	0.0	0.0	0.0	0.0	0.0
.07--<.10	0.0	0.0	0.0	0.0	0.0	0.0	0.0	0.0	0.0	0.0
.10--<.15	0.0	0.0	0.0	0.0	0.0	0.0	0.0	0.0	0.0	0.0
.15--<.20	0.0	0.0	0.0	0.0	0.0	0.0	0.0	0.0	0.0	0.0
.20--<.25	0.5	0.2	0.1	0.1	0.1	0.2	0.3	0.6	0.1	0.3
.25--<.30	0.4	0.2	0.1	0.1	0.1	0.3	0.3	0.5	0.9	0.2
.30--<.35	0.1	0.0	0.0	0.0	0.0	0.0	0.1	0.1	0.2	0.1
.35--<.40	0.2	0.1	0.1	0.1	0.0	0.1	0.1	0.3	0.5	0.1
.40--<.45	0.1	0.0	0.0	0.0	0.0	0.1	0.1	0.1	0.2	0.1
.45--<.50	10.6	5.1	6.4	5.5	0.5	1.3	1.9	3.2	19.3	4.6
.50--<.60	6.5	3.2	4.3	3.7	0.2	0.7	1.5	2.6	12.6	3.0
.60--<.70	1.1	0.6	5.2	4.5	0.2	0.4	0.2	0.4	6.8	1.6
.70--<.80	0.8	0.4	1.1	1.0	0.2	0.4	0.6	1.0	2.7	0.6
.80--<.90	0.3	0.2	0.4	0.3	0.1	0.3	0.2	0.3	0.9	0.2
.90--1.00	185.7	90.0	97.9	84.7	36.1	96.1	52.8	90.9	372.4	89.2
Total	206.3	100.0	115.6	100.0	37.5	99.9	58.1	100.0	416.6	100.0

SOURCE: 1982 NRI.

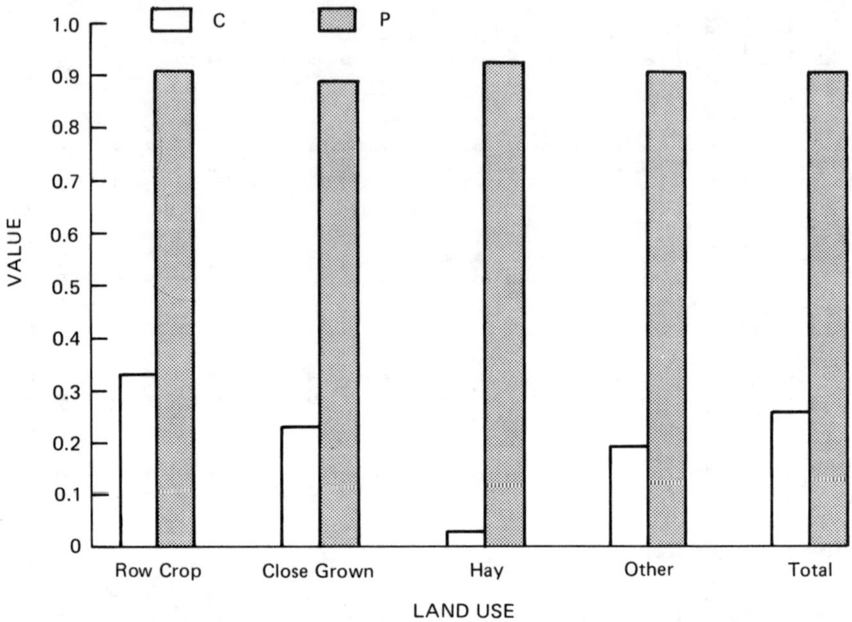

FIGURE 3 Weighted average values of C and P factors by land use in the United States, 1982.

SOURCE: 1982 NRI.

Contour farming does not directly affect the C-factor value, but nevertheless has a similar distribution to minimum tillage (see Figure 4). The practice of contour stripcropping does affect the value of the C factor (through planting close-grown crops and meadow in rotation), but the modal mean is representative of a broader range, from .07 to less than .15. These values are significantly lower than minimum tillage, and are likely due to the impact of meadow and close-grown crops.

Terrace systems do not directly affect the C-factor value. They have about the same modal mean and distribution spread as minimum tillage and contour curves (see Figure 4).

The distribution for minimum tillage, contour farming, stripcropping, and terraces by P factor is shown in Table 10. Minimum tillage has a small distribution around the typical P-factor values of .45 to less than .60 (see Figure 5). Over 84 percent of minimum tillage points

TABLE 8 Acreage and Distribution of Selected Conservation Practices in the United States, 1982

Conservation Practice	Acres (Million)	Distribution (Percent)	Percent Use	
			Row Crops	All Cropland
Minimum tillage	100.2	60.0	48.6	24.0
Contour system	34.9	20.9	16.9	8.4
Stripcropping	3.4	2.0	1.6	0.8
Terrace system	28.5	17.1	13.8	6.8
Total	167.0[a]	100.0	80.9	40.0

[a]Double counting is present in total. Each sample point could have up to three practices. Therefore, actual acreage is between 100.2 and 167.0 million acres.

SOURCE: 1982 NRI.

fall in the .90 to 1.0 category. This would indicate that minimum tillage does not tend to occur in conjunction with practices that affect P-factor values. Contouring and terracing, conversely, have similar distributions and tend to have a modal mean in the .45 to less than .60 categories (see Figure 5). Stripcropping has a modal mean in the .20 to less than .30 categories, the lowest of all practices (see Table 10). Smaller peaks also occur in the .35 to less than .40, .45 to less than .50, and .90 to 1.00 categories, indicating that stripcropping does occur with contour and terraces.

The record of P-factor values of .90 and above indicates that contours and stripcropping occur on lands that have slopes or slope lengths that all but eliminate the impacts of the supporting practices. To some extent, these phenomena cannot be prevented due to the use of a variety of slopes and slope lengths in close proximity.

Terraces were built on 28.5 million acres, about 7 percent of the cropland. The value of the P factor was in the range of .45 to less than .60 some 68 percent of the time. P-factor values of .90 to 1.00 occurred nearly 24 percent of the time on terraced acres (see Table 10). The distribution of this acreage by RKLS group is shown in Table 11. Terraces occur on all the RKLS groups studied, with a broad base from 0 to less than 75 tons/acre/year and some concentration at 5 to less than 25 tons/acre/year. Terraces appear to reduce slope length by about 100 feet in most RKLS groups (see Table 11).

TABLE 9 Acreage and Distribution of Cropland Uses by C Factor, United States, 1982

C Factor	Minimum Tillage Acres (Million)	Minimum Tillage Percent	Contour Farming Acres (Million)	Contour Farming Percent	Stripcropping Acres (Million)	Stripcropping Percent	Terracing Acres (Million)	Terracing Percent
.00--<.02	1.5	1.46	0.6	1.78	0.2	6.83	0.6	2.17
.02--<.05	2.4	2.43	0.8	2.22	0.3	9.09	0.4	1.40
.05--<.07	1.7	1.73	0.6	1.62	0.3	8.25	0.2	0.76
.07--<.10	6.8	6.75	2.1	6.10	0.9	25.81	1.2	4.26
.10--<.15	17.6	17.61	5.5	15.79	0.8	22.85	5.2	18.11
.15--<.20	21.7	21.62	6.2	17.77	0.3	7.81	5.9	20.78
.20--<.25	15.8	15.77	4.3	12.18	0.2	5.39	4.5	15.88
.25--<.30	10.9	10.88	2.9	8.25	0.1	4.31	2.2	7.70
.30--<.35	10.7	10.68	3.1	8.74	0.1	2.19	1.9	6.73
.35--<.40	5.4	5.38	3.3	9.58	0.1	3.96	2.3	7.90
.40--<.45	2.4	2.43	1.4	4.15	0.1	1.52	1.0	3.61
.45--<.50	1.7	1.67	1.5	4.33	0.0	0.84	0.8	2.73
.50--<.60	1.2	1.15	0.9	2.62	0.0	1.01	0.8	2.90
.60--<.70	0.4	0.37	1.6	4.71	0.0	0.15	1.4	4.88
.70--<.80	0.0	0.02	0.1	0.16	0.0	0.00	0.1	0.20
.80--<.90	0.0	0.02	0.0	0.00	0.0	0.00	0.0	0.00
.90--1.00	0.0	0.03	0.0	0.00	0.0	0.00	0.0	0.00
Total	100.2	100.00	34.9	100.00	3.4	100.01	28.5	100.01

SOURCE: 1982 NRI.

Row crops are planted on 41 percent of terraced land, compared with 50 percent on all cropland (see Table 12). The largest use of terraced land is for close-grown crops (45 percent, versus 28 percent of all cropland). Hay acreage accounts for 2 percent of terraced cropland, a drop from 9 percent of total cropland. "Other cropland" changed the least, with a drop from 14 percent of all land to 11 percent of terraced land. The distribution of C values is not significantly different from that of minimum tillage and contour farming (see Table 9).

For land in row crops, small grains, and hay, average RKLS values are higher for terraced cropland. For the other cropland category, however, the RKLS value is lower on terraced cropland (see Tables 12, 13, and 14). This indicates that for the major use categories, terraces were installed on the land with the highest potential for erosion.

The relationship between erosion and slope lengths greater than 90 feet is shown in Table 15. About 60 percent of the acreage has slope lengths of 200 feet or

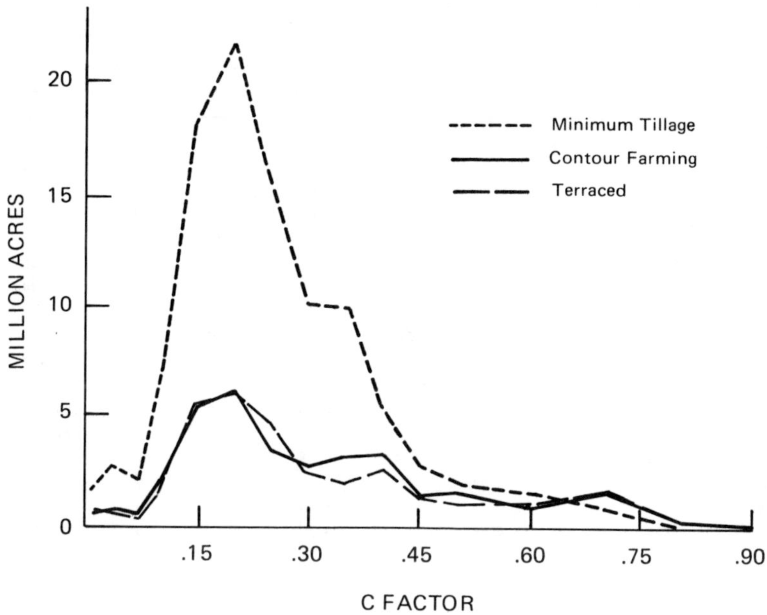

FIGURE 4 Distribution of minimum tillage, contour farming, and terraced acreage by C factor in the United States, 1982.

SOURCE: 1982 NRI.

less. The next largest concentration is in the over-350-feet category. The distribution of soil movement has the same pattern: On a per acre basis, the weighted average tons/acre generally increases along with slope length, until the 351-feet-and- larger category, where the C factor reduces soil movement more than slope length increases it.

If all slope lengths were set at 90 feet, the total soil movement would be reduced by 2.5 billion tons, and 7 fewer tons/acre would be moved (see Table 15). This would require adding terraces to 343 million acres. If average terrace costs were $300/acre, the cost would be $103 billion--or $41/ton. At an average ton/acre reduction of 7 tons, the average cost would be $287/acre. Unfortunately, the tons/acre loss would still be two to five times above tolerable levels.

TABLE 10 Acreage and Distribution of Selected Conservation Practices by P Factor, United States, 1982

P Factor	Minimum Tillage Acres (Million)	Minimum Tillage Percent	Contour Farming Acres (Million)	Contour Farming Percent	Stripcropping Acres (Million)	Stripcropping Percent	Terracing Acres (Million)	Terracing Percent
.00--<.02	0.00	0.00	0.00	0.00	0.00	0.00	0.00	0.00
.02--<.05	0.00	0.00	0.00	0.00	0.00	0.00	0.00	0.00
.05--<.07	0.00	0.00	0.00	0.00	0.00	0.00	0.00	0.00
.07--<.10	0.00	0.00	0.00	0.00	0.00	0.00	0.00	0.00
.10--<.15	0.00	0.00	0.00	0.00	0.00	0.00	0.00	0.00
.15--<.20	0.00	0.00	0.00	0.00	0.00	0.00	0.00	0.00
.20--<.25	0.21	0.21	0.18	0.53	0.85	25.13	0.08	0.28
.25--<.30	0.19	0.19	0.17	0.48	0.68	20.18	0.05	0.19
.30--<.35	0.02	0.02	0.10	0.06	0.13	3.95	0.01	0.02
.35--<.40	0.09	0.09	0.10	0.28	0.37	10.92	0.03	0.11
.40--<.45	0.04	0.04	0.03	0.08	0.18	5.16	0.01	0.04
.45--<.50	6.18	6.16	17.40	49.85	0.40	11.74	11.76	41.23
.50--<.60	4.88	4.86	11.19	32.04	0.24	7.16	7.50	26.34
.60--<.70	2.98	2.98	0.44	1.27	0.05	1.48	1.90	6.83
.70--<.80	0.85	0.85	0.94	2.69	0.05	1.47	0.25	0.88
.80--<.90	0.18	0.18	0.35	1.01	0.04	1.20	0.03	0.10
.90--1.00	84.59	84.42	4.09	11.71	0.39	11.61	6.84	23.98
Total	100.21	100.00	34.99	100.00	3.38	100.00	28.46	100.00

SOURCE: 1982 NRI.

When slopes over 90 feet are sorted by slope (*see* Table 16), group A (a slope of 0 to less than 2 percent) accounts for 60 percent of the acres. Another 24 percent of the acreage can be found in slope group B (2 to less than 5 percent slope) and 10 percent is in slope group C (a slope of 5 to less than 9 percent). The distribution curve drops down to less than 1 percent for slope group G (25 percent slope or higher). Slope groups A to D (0 to less than 14 percent slopes) account for 82 percent of the tonnage of soil movement. Within any one slope length, weighted averages of tons moved per acre rise sharply as slope increases. Within slope groups, a dichotomy exists when slope lengths are increased. Slope groups A and B have ranges from low to high that are within 3 tons/acre of their respective weighted averages (*see* Table 16). Slope group C has a range of 18

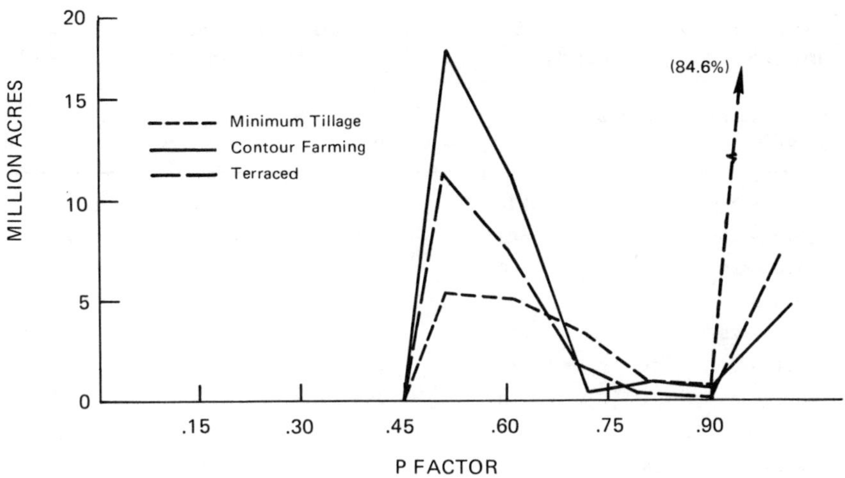

FIGURE 5 Distribution of minimum tillage, contour farming, and terraced acreage by P factor in the United States, 1982.

SOURCE: 1982 NRI.

tons/acre, but the lowest and highest figures are still within 9 tons/acre of the weighted average. Slope groups higher than C have progressively larger ranges.

The implications are that flat soils have a full range of slope lengths. Soils with very low RKLS (less than 5 tons/acre/year) may account for a significant proportion--more than one-third of the acreage--of land with slope lengths greater than 350 feet.* The volatile combination

*Many researchers are proposing using $RKLST^{-1}$ as a statistic to assign fragile ranking to soils. Soils where RKLS values are less than 5 are ranked as nonfragile. One-third of the soils with less than 2 percent slope will have RKLS values of less than 5 tons/acre. If the soil loss tolerance limit is 4 or 5 tons/acre, then $RKLST^{-1}$ will be less than 1 and yet the soil may be experiencing twice the tolerable amount of soil movement. Testing the magnitude of the potential error is beyond this paper. The subject is being addressed in another report by the authors and will be available in the future.

TABLE 11 Amount and Slope Length of Terraced and Unterraced Cropland, by RKLS Factor, United States, 1982

RKLS (Tons/ Acre/Year)	Slope Length (Feet) Terraced	Unterraced	Total	Acres (Million) Terraced	Unterraced
0.0--<5	192	340	338	1.4	112.9
5.0--<10	208	303	297	6.6	98.2
10.0--<15	185	280	272	4.9	53.1
15.0--<20	175	273	262	4.0	30.1
20.0--<25	155	276	259	3.0	17.9
25.0--<30	146	264	247	2.2	12.6
30.0--<35	151	253	238	1.5	9.1
35.0--<40	143	241	229	0.9	6.9
40.0--<50	138	235	224	1.3	10.2
50.0--<75	148	237	229	1.5	15.1
75.0--<100	149	235	230	0.6	8.3
100.0 & up	188	234	232	0.7	14.8
Net average	175	295	287		

SOURCE: 1982 NRI.

of high slope length and high slopes can result in very high erosion rates. Fortunately, as noted earlier, cropland acreage is skewed to the lower slope groups, where soil movement can be more easily controlled. It would also appear that change in slope length and change in slope are indirectly proportional, i.e., when one goes up, the other one goes down and vice versa.

Overall Comparison

The interaction of supporting practices, land use, C and P factors, and RKLS groups is established by sorting the NRI cropland data into those points with one or all supporting practices (minimum tillage, contouring, strip-cropping, or terracing) and those points without any such practices.

A comparison of C-factor values by land use, RKLS, and supporting practices shows that land in row and close-grown crops has slight decreases in C values as RKLS increases from 0 to 100 or more tons/acre/year (see Figure 6). The C-factor values for hayland are not

TABLE 12 Acreage and RKLS Factor on Terraced and Unterraced Land, by Land Use, United States, 1982

Land Use	Terraced Land Acres (Million)	Terraced Land RKLS (Tons/ Acre/Year)	Unterraced Land Acres (Million)	Unterraced Land RKLS (Tons/ Acre/Year)
Row crops	11.8	30.6	194.5	21.7
Close-grown crops	12.8	19.5	102.8	14.9
Hay	0.7	40.2	36.8	35.1
Other crops	3.2	20.8	55.0	24.7
Total	28.5	24.8	389.1	21.6

SOURCE: 1982 NRI.

affected by rising RKLS values. Land in other crops is similar to close-grown crops for low RKLS values, but the C factor values are about half as high when the RKLS is high.

Nationally, the impact of supporting practices on row crops causes the weighted average C-factor value to fall from 0.35 to 0.28 (see Figure 7). The two lines have almost parallel decreases, with the exception of the lowest RKLS values (less than 10 tons/acre/year). Overall, both the with and without trend lines decreased about .08 points from RKLS values of less than 5 tons/acre/year to those over 100 tons/acre/year.

The change in weighted average C-factor values for close-grown crops is from .26 for without supporting practices to .20 with them (see Figure 8). This difference of .06 is maintained throughout the lower range of values. The upper range of RKLS values has a .07 difference. The trend lines are similar to those for row crops, except that here the C-factor values decrease faster with supporting practices than without them. The overall decreases were .056 with these practices and .042 without them. As with row crops, close-grown crops show more variability in the lowest RKLS groups.

The change in weighted average C-factor values for hayland is from .04 with supporting practices to .03 without (see Figure 8). The with-supporting-practice trend line is almost flat while the without line declines from .03 to .02 as RKLS ranges from 0 to greater than 100 tons/acre/year.

TABLE 13 Sheet and Rill Erosion on Terraced Cropland by Land Use and RKLS Factor, United States, 1982 (Million Tons)

RKLS (Tons/Acre/Year)	Row Crops	Close-Grown Crops	Hay	Other Crops	Total
0--<5	2.6	1.8	0.1	0.7	5.3
5--<10	16.1	24.2	0.5	8.8	49.6
10--<15	18.1	33.6	0.7	7.8	60.2
15--<20	26.8	35.5	1.5	5.3	69.1
20--<25	26.5	34.1	1.8	4.2	66.5
25--<30	26.1	27.7	2.1	4.3	60.3
30--<35	22.4	20.4	1.6	3.0	47.3
35--<40	18.1	13.1	1.2	3.3	35.7
40--<50	35.6	17.6	2.5	3.9	59.5
50--<75	56.3	20.3	5.5	7.1	89.4
75--<100	34.4	7.2	3.4	4.3	49.3
100 & up	77.7	15.2	8.0	13.6	114.5
Total	360.7	250.7	28.9	66.3	706.7

TABLE 14 Sheet and Rill Erosion on Unterraced Cropland by Land Use and RKLS Factor, United States, 1982 (Million Tons)

RKLS (Tons/Acre/Year)	Row Crops	Close-Grown Crops	Hay	Other Crops	Total
0--<5	137.3	106.1	26.3	51.3	321.1
5--<10	398.6	180.3	47.8	89.9	716.6
10--<15	370.5	170.1	36.8	75.4	652.8
15--<20	286.2	125.7	45.3	65.2	522.5
20--<25	214.3	102.5	33.6	49.5	399.9
25--<30	173.8	86.2	38.0	46.2	344.1
30--<35	152.9	72.7	30.0	39.3	294.8
35--<40	132.1	56.8	34.7	34.9	258.4
40--<50	235.1	97.9	59.4	63.0	455.4
50--<75	485.3	173.2	125.3	134.8	918.7
75--<100	367.8	110.6	125.9	104.7	709.1
100 & up	1,285.7	251.2	690.5	607.5	2,834.9
Total	4,239.6	1,533.3	1,293.6	1,361.7	8,428.3

SOURCE: 1982 NRI.

TABLE 15 Cropland, Current Soil Loss, and Soil Loss If Slope Lengths Reduced to 90 Feet, by Slope Length, United States, 1982

Slope Length (Feet)	Acres (Million)	Soil Loss (Billion Tons) Current	Reduced	Difference	Soil Loss (Tons/Acre) Current	Reduced	Difference
91--150	111.7	2.4	2.1	0.3	22	19	3
151--200	62.0	1.7	1.2	0.5	27	19	8
201--250	23.6	0.8	0.5	0.3	34	22	12
251--300	41.3	1.1	0.7	0.4	28	16	11
301--350	10.0	0.4	0.2	0.2	35	19	16
351 & up	94.3	1.6	0.8	0.8	18	9	9
Total	342.9	8.0	5.5	2.5	--	--	--
Weighted average	--	--	--	--	23	16	7

SOURCE: 1982 NRI.

The change in weighted average C-factor values for other cropland is from .19 with supporting practices to .20 without them (see Figure 7). Except for the lowest RKLS values, the two trend lines are not significantly different. Like the other land uses, most of the variability and volatility was for land with an RKLS less than 20 tons/acre/year.

A comparison of P-factor values by land uses, RKLS values, and supporting practices is much more varied (compare Figures 6 and 9), particularly in the lower RKLS groupings. It is not until RKLS increases to 40 tons/acre/year that separate trends develop. All land use trend lines for P when supporting practices are present decline more or less together through RKLS of 30 to 40 tons/acre/year. Land in hay and "other crops" continue to exhibit similar trends. Row and close-grown crops also exhibit similar trends except for the RKLS range of 25 to 90 tons/acre/year. At these values, close-grown crops stop having lower P values than row crops and stabilize to have higher P values.

The weighted average P-factor values for each land use are as follows: for land in row crops, .86 with supporting practices and 1.0 without; for close-grown crops, .87 with and .98 without; for hayland, .81 with and 1.0 without; and for land in other crops, .88 with supporting practices and .99 without them.

TABLE 16 Comparison of Ranges and Weighted Averages of Soil Loss Per Acre by Slope Group, United States, 1982

Slope Group (Percent of Slope)	Tons/Acre Lowest	Highest	Range	Weighted Average
A (0--<2)	7	10	3	8
B (2--<5)	20	25	5	22
C (5--<9)	44	62	18	53
D (9--<14)	90	128	38	106
E (14--<18)	135	176	41	151
F (18--<25)	151	245	94	194
G (25 or more)	105	490	385	391

SOURCE: 1982 NRI.

SUMMARY AND CONCLUSIONS

For decades the U.S. agricultural sector has produced more than was demanded domestically or than could be sold abroad at a profit. As a result, consumers have benefited from low food and fiber prices while the agricultural sector has suffered from high production costs relative to low prices and declining net farm income. As farm programs have sought to control acreages and reduce the cropland base, farmers have turned to intensification and increased production as their only means to solve their situation. Because prospects for the rest of the 1980s suggest that supply will probably continue to exceed demand, such intensive cropland use will probably also continue with the accompanying effect of soil erosion.

Using 1982 NRI data, this paper explores the extent and severity of sheet and rill erosion problems, the extent to which conservation practices have reduced potential erosion, and the ability of conservation practices to resolve sheet and rill erosion problems.

The NRI data indicate that 417.5 million acres are readily available for crop production in the United States. Soil movement per acre from sheet and rill erosion is 4.3 tons/acre/year overall. Even cropland used for row-crop production has an erosion rate of 6.1

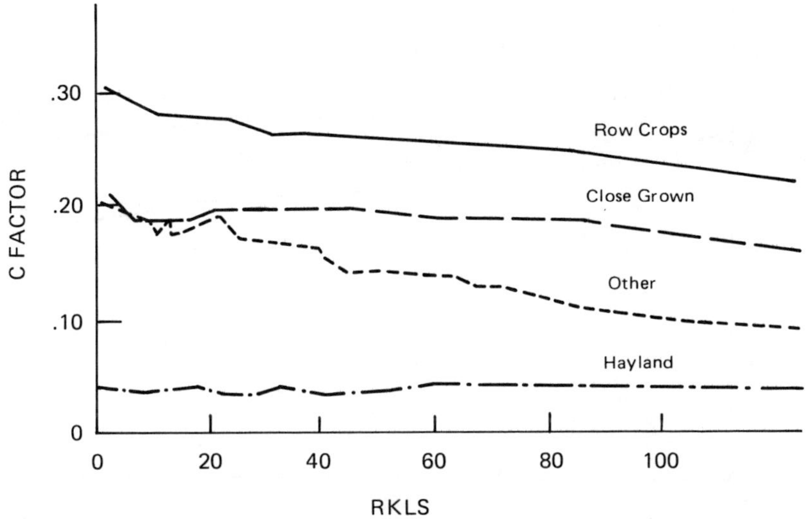

FIGURE 6 Comparison of C-factor values for selected land uses by RKLS groups in the United States, 1982.

SOURCE: 1982 NRI.

tons/acre/year. Row crops are planted on almost one-half of the cropland acreage but have almost 70 percent of sheet and rill erosion. Over 78 percent of all national cropland has sheet and rill erosion of less than 5 tons/acre/year, 90 percent of cropland has 10 tons/acre/year, and almost 95 percent of cropland has 15 tons/acre/year. Crop production occurs on almost 87 percent of soils.

The natural potential for sheet and rill erosion is also revealed by the NRI sample. The natural potential for erosion of cropland is calculated to be about 21.8 tons/acre/year. This suggests that if the 417.5 million acres used for cropland had been in fallow and had been tilled up and down the slopes, the average annual erosion rate on these lands would have been about 21.8 tons/acre/year. Because these factors were not present, erosion was much less severe and averaged only 4.3 tons/acre/year. The difference is due to the impact of current managerial decisions regarding cover and management practices, which are reflected in the C factor of the soil loss equation; to supporting practices, which are reflected in the P factor; and to shortened slope lengths in terraces.

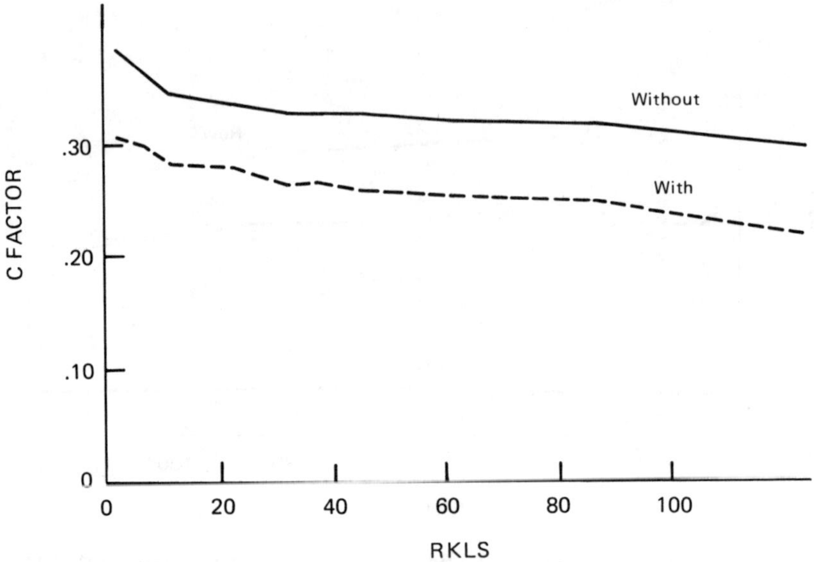

FIGURE 7 Comparison of C-factor values for land in row crops and other crops with and without supporting practices, by RKLS value, in the United States, 1982.

SOURCE: 1982 NRI.

A comparison of the average tons per acre sheet and rill erosion for various croplands shows that row crops are grown on lands that are nearly average for all cropland. Close-grown crops are grown on the least erosive land, and hay is grown on the most erosive land.

A comparison of C-factor values across land uses indicates that plant matter and residue are the most effective in reducing erosion rates on hayland and the least effective on cropland. This is not surprising as the overall impact of hay is to lower the potential erosion value of 125 tons/acre/year to 5 tons/acre/year.

About one-third of U.S. land has sheet and rill erosion rates greater than 15 tons/acre/year. These soils have one-third of total erosion but constitute only 6 percent of the total cropland. Thus, if these 25 million acres could be taken out of cultivation, excess erosion could be decreased by almost one-third.

Another way to lower soil erosion is through supporting practices that lower the value of the P

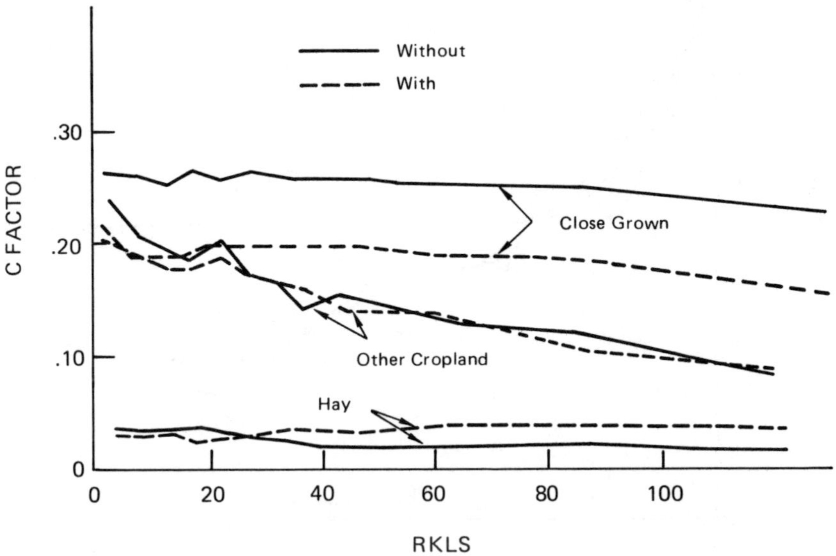

FIGURE 8 Comparison of C-factor values for land in close-grown crops, other cropland, and hay with and without supporting practices, by RKLS value, in the United States, 1982.

SOURCE: 1982 NRI.

factor. The weighted average value of the P factor reported in the 1982 NRI for all cropland was .91. Thus, the overall impact of supporting practices was to reduce sheet and rill erosion by 9 percent--a reduction of about 2 tons/acre. One reason for this small impact is that only 40 percent of cropland has a supporting practice that lowers the P-factor value.

From a national policy viewpoint, it appears that land use changes reflecting plant and residue changes are more important than supporting practices. This point has received little attention because comparisons are often made between supporting practices, given plant and residue cover. Many other studies compare various systems of plant and residue management without considering the alternative of support practices.

Other ways to lower erosion rates are contour farming, strip-cropping, and terracing. There are about 167 million acres in some form of minimum tillage, contour

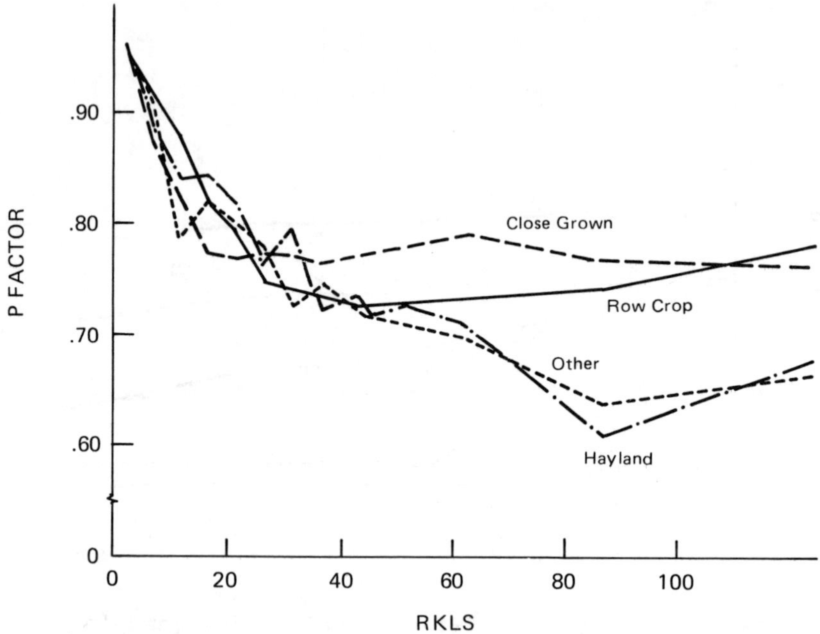

FIGURE 9 Comparison of P-factor values for selected land uses, by RKLS groups, United States, 1982.

SOURCE: 1982 NRI.

systems, stripcropping, and terrace systems in the United States, including 81 percent of row crops and 40 percent of all cropland. Minimum tillage, the dominant conservation practice used, is concentrated on 49 percent of row-crop acres. Contouring and terrace systems rank a distant second and third.

For some cropland uses, terracing reduces erosion. However, for the major land uses, terraced land has a higher potential for erosion than unterraced land. Values that reflect the potential for sheet and rill erosion are higher for terraced cropland than unterraced when land is used for row crops, small grains, and hay. For other cropland uses, terraced cropland has a lower potential for erosion.

Erosion can also be reduced by changing slope lengths of cropland. The impact of adjusting slope lengths greater than 90 feet to 90 feet, which would happen with terracing, would be to reduce total tons of soil movement by 2.5 billion tons and the tons/acre by 7 tons. This would require adding terraces to 343 million acres at an estimated average cost of $287/acre. Unfortunately, the tons/acre loss would still exceed tolerable levels by two to five times.

An additional study is needed that extends the methods used in this paper to MLRAs, or to a selection of MLRAs across the United States. Such a study would reveal the regional impacts of erosion control practices and regional solutions to natural resource problems. A more detailed regional and MLRA study would have different results from this national study.

Discussion

Arnold R. Miller

Rosenberry and English properly approach the soil erosion problem by recognizing that it is a physical process that occurs within the context of and is inextricably tied to economic forces.

The engine of production that combines natural, human, capital, and technologic resources to produce agricultural outputs is the farmer. If our concern is for the optimum preservation of soil resources in a market economy, a basic fact that must be recognized and acted upon is that the value of soil in farm production is derived from the value of the commodities it is used to produce. In a market economy, every force that operates to limit the private and social value of farm outputs also operates to limit the value of soil and the volume of soil conservation that is privately and socially justifiable.

Forces affecting the value of soil in farm production bear on the erosion process through their impact on the farmer. They do so by influencing the farmer's ability to convert production inputs of given value to outputs of greater value over successive cycles of production.

Five basic forces interact to determine the value of a unit of erosive soil in farm production at any point in time: (1) the price of farm outputs; (2) the change in output per unit of soil loss; (3) the interest rate; (4) the rate of change in the price of farm outputs; and (5) the rate of technologic advance as it applies to farm production.

The value of erosive soil to the farmer operating in a market economy rises and falls with the value of the commodities it is used to produce. This elementary relationship may not warrant discussion, but too often this relationship is ignored. The link between the value of farm outputs and the production value of soil was ignored for decades in the administration of federal conservation programs. Large portions of available funds and staff resources earmarked for conservation were diverted to production-oriented practices, especially drainage and irrigation. The cumulative effect of these diversions was to increase production on millions of acres. The increased production aggravated crop surpluses and increased downward pressure on farm prices and on the value of soil in farm production.

As for crop yield response to soil loss, three general outcomes are possible: (1) yields may not change in response to soil loss, at least not at present soil depths; (2) yields may decline in response to soil loss; and (3) yields may rise in response to soil loss, such as when underlying soil material provides a better rooting medium than surface material.

In situations where crop yields do not decline in response to soil loss, one has to question what is achieved in the way of preserving production capacity by expending resources to reduce erosion. Where yields decline in response to soil loss, one must distinguish between soil deterioration and soil depletion. Following Schickele (1937) and Bunce (1942), soil deterioration refers to the permanent impairment of the ability of soil to support plant growth. Soil depletion refers to the removal of plant nutrients and organic matter by any means when they can be replaced and fertility restored. Soil deterioration involves the permanent loss of productive capacity in the sense of consuming an exhaustible resource. By contrast, soil depletion involves the sacrifice of renewable productive capacity.

The literature suggests that where crop yields decline in response to erosion, the range of decline is from 1 to 9 percent per acre inch of soil loss and the average decline is about 5 percent. Assuming corn yields of 120 bushels/acre and soil weight of 150 tons/acre inch, 1 ton of soil loss coverts to a yield reduction of 0.04 bushel annually. For lack of a better term, this value will be called the annual yield equivalent of a ton of soil loss. When the annual yield equivalent of a ton of soil loss is multiplied by crop price at the farm gate, an estimate of the annual decline in gross income resulting from ton of soil loss is obtained.

The annual decline in gross income due to erosion can be capitalized to estimate the value of erosive soil in agricultural production. The techniques used in the capitalization process are widely used in the private sector and elsewhere when the objective is to obtain maximum performance over time from available productive resources.

The interest rate is a key item in the capitalization process. Experience suggests that parts of the conventional wisdom of soil conservation view the interest rate as a financial ogre feeding on the flesh of unborn generations by discounting future food supplies to present values. The emotional appeal of this view is strong. Its error, however, is equally strong and operates to reduce future resource endowments rather than enhance them.

Few will argue that parents and grandparents do not serve their descendents best by seeking the highest annual yield on investments set aside for them. Although the relationship between current investment yield and future resource endowments is most commonly thought of in terms of financial assets, it holds equally well for physical assets, such as soil. When investment yields are reinvested, they compound over successive periods. Thus, by demanding high performance from investments set aside for the future, unborn generations stand to benefit from larger rather than smaller resource endowments.

Although not always intuitively obvious, the arithmetic of the situation demonstrates that the effect of discounting is simply to reverse the effect of compounding. Thus, insofar as we accept low rates of return on investments in soil preservation, we not only retard their current performance but also reduce future resource endowments. A major impact of applying artificially low interest rates to soil conservation practices and programs, or any program, is to penalize future resource endowments in favor of poor current performance. I submit that insofar as the purpose of soil conservation is to ensure maximum availability of resources for future generations, conservation programs and practices should be planned and administered using interest rates at least as high private market rates.

The annual rate of change in the price of farm outputs is also central to the valuation of erosive soil. Insofar as prices received by farmers can be expected to rise at a given rate, the return on conservation investments will increase. The easiest way to integrate expected price increases properly into the valuation process is to subtract the annual rate of price increase from the interest rate. Conversely, the easiest way to integrate expected price declines properly into the valuation process is to add the annualized rate of price decline to the interest rate.

The relationship between the rate of change in farm prices and the interest rate is critically important to soil conversation in a market economy. Half a century ago, Harold Hotelling (1931) demonstrated that the value of exhaustible resources must rise at an annual rate equal to the interest rate if the privately and socially optimum rates of exhaustion are to coincide.

Applying Hotelling's "law" to agriculture, if the price of farm outputs rises at annual rates equal to the interest rate, market forces will give farmers every incentive to conserve erosive soil to the extent socially justifiable. The sobering secular trend, however, is for inflation-adjusted farm prices to decline rather than to rise, and therein lies the primary cause of the erosion problem. A "cheap food" policy manifests itself as a "cheap resource" policy where erosive soil and farm production in a market economy are concerned.

Fortunately, technical advance can have the same effect as rising farm prices in determining the agricultural production value of erosive soil. Where technical advance is "neutral" with respect to soil, the benefits of advancing technology are realized without expenditure of resources to "adapt" soil to successively higher levels of technology. When expressed as an annual rate of change in output per acre, technical advance is properly integrated into the soil valuation process by subtracting the rate of advance from the interest rate.

Conversely, technical decline is properly integrated into the valuation process by adding the rate of decline to the interest rate. Other things being equal, higher rates of technical advance translate to higher unit values of erosive soil. Hence, research to speed the rate of technical advance in agriculture can act to offset the exploitive impact of a "cheap food" policy on erosive soil resources.

The determinants of the value of erosive soil in agricultural production interact to present the farmer figuratively with a basic question as to the effectiveness of erosion control practices. Are erosion control practices efficient enough to reduce soil loss at costs that are less than the capitalized value of the soil assets they preserve? Casual empiricism suggest that under recent conditions, the capitalized value of erosive soil in agricultural production ranges from a few cents to a few dollars per ton. Studies indicate that the cost of erosion control can range from almost nothing to many dollars per ton. These observations suggest that erosion control practices can be applied selectively in such manner as to be both "effective" as measured in tons/acre of erosion reduction and "efficient" in the sense of expending resources of lesser value than that of the soil they preserve.

At risk of over-simplification, one can observe the history of federal erosion control programs as a series of phases. Without attempting to attach dates to phases, it is safe to characterize the earliest phase as having a tremendous zeal for the application of conservation practices coupled with Depression-era make-work programs. The result was the application of erosion control practices almost without regard to the scarcity of resources other than soil itself.

The second phase can be characterized as striving for effectiveness in controlling erosion. Unfortunately, effectiveness was defined in the absolute terms of the soil loss tolerance, or T value. This was an improvement over the first phase, however, in that the T value concept recognizes that there may be situations where the expenditure of scarce resources to reduce erosion may not be justifiable. Great progress was made during the second phase in terms of technical capability to quantify erosion problems and the impact of alternative erosion control practices on them both at the farm level and in terms of national policy. Too often, however, the drive to reduce erosion rates to absolute T values established without regard to nonsoil resources caused farmers to question whether the federal government really knew much about "scarce resources."

It is time for a third phase in the erosion control effort. The third phase should strive for balance, in the sense of recognizing the scarcity of all types of resources, natural and otherwise. Balance means two things with respect to erosion control practices, effectiveness in reducing erosion and efficiency in both getting the job done and in determining how far to go.

REFERENCES

Bunce, A. C. 1942. Economics of Soil Conservation. Ames: Iowa State.

Hotelling, H. 1931. The economics of exhaustible resources. Journal of Political Economy 39:137-75.

Schickele, R. 1937. Economics of Agricultural Land Use Adjustment. 1. Methodology in Soil Conservation and Agricultural Adjustment Research. Research Bulletin 209. Iowa Agricultural Experiment Station.

8
Applications of the NRI Data to Inventory, Monitor, and Appraise Offsite Erosion Damage

Lee A. Christensen

There are two general impacts of soil erosion from agricultural land. Onsite effects, those occurring at the field or farm level, are primarily reflected in soil productivity changes associated with erosion. Offsite impacts occur primarily when soil and chemicals are carried from fields and farms in runoff, causing water pollution and deposition problems downstream and groundwater infiltration. This paper addresses the applicability and use of the U.S. Department of Agriculture (USDA) National Resources Inventory (NRI) to investigate such offsite damages.

Historically, public attention and funds have concentrated on reducing the adverse onsite impacts of soil erosion. Increased concerns about water quality degradation and the associated cleanup costs, combined with passage of clean water legislation during the 1970s, focused attention on offsite impacts. Negative offsite impacts have broadened public concern about soil erosion to include more than soil productivity issues. Soil leaving a field due to water erosion represents costs not only to the farmer and consumer in terms of lost profits and high food prices, but also to those downstream, for associated cleanup costs. Solving offsite problems has been complicated by the lack of information and capabilities to explain the diffuse and complex nature of the physical relationships. Recent studies in this area have vastly improved the ability to explain and model the physical processes (Bailey and Swank, 1983). This work is being linked with economic data to develop more complete assessments of the offsite damages and the alternatives for their control and reduction.

The NRI conducted by the Soil Conservation Service (SCS) in 1977 and 1982 is a data source that might help assess offsite damages associated with soil erosion (USDA, 1984). This paper considers such a use of the NRI data. It begins with a brief discussion of terms and the nature of offsite pollution, followed by an examination of some applications of the NRI data. Possible uses of the data in conjunction with available water quality models to address water quality questions are explored. The final section focuses on potential uses of the NRI and presents some suggestions for future inventories.

THE NATURE OF OFFSITE DAMAGES

Water bodies receive pollution loads from point sources, such as municipalities and industrial plants, and nonpoint sources, including agriculture. Nonpoint source pollution originates from ill-defined and diffused sources, such as urban areas, cultivated fields, forests, and pastures. Most agricultural nonpoint source pollution is caused by sediment and sediment-transported chemicals (Bailey and Waddell, 1979). It does not include runoff from urban areas, mining and construction activities, highways, logging activities, or streambank erosion.

The offsite effects of agricultural nonpoint source pollution are diverse and complex. Each type of pollutant has unique characteristics, both with respect to the mode of transport through the water course, and the fate of the pollutant as it moves from field to stream to lake to river or reservoir. Heavy sediment loads can fill reservoirs and cause channel siltation, which raises the costs of water treatment and channel dredging. Excess levels of nitrogen or phosphorus in streams, lakes, or estuaries can cause eutrophication. Sediment and chemicals can have adverse effects on fish and wildlife, greatly reducing the economic and recreational value of streams and lakes.

It is important to understand the distinction between problems with surface water quality and those related to groundwater or contamination. Agriculturally related groundwater questions arise primarily with regard to the leaching of nitrates and to soluble persistent pesticides. Solutions to surface water quality problems are not necessarily answers to groundwater problems.

APPLICABILITY OF NRI DATA TO WATER QUALITY ISSUES

Design Constraints

The NRI is a tremendously rich source of information, but with serious limitations for addressing water quality questions. It is designed to systematically develop information on the condition of the nation's agricultural land base every 5 years. Its area of primary application is the onsite, or farm, level. The NRI data base can be used to address some of the water quality issues influenced by activities based on agricultural land, primarily sediment loadings. However, it contains no direct data for the analysis of water quality problems attributable to nonagricultural sources.

Levels of Detail and Aggregation

Analysis of water quality problems due to agricultural sediment is facilitated if data can be aggregated along hydrologic boundaries, such as a watershed or river basin, rather than Major Land Resource Area (MLRA) boundaries. However, since the NRI is considered reliable at the MLRA level, information for small watersheds that are fractions of MLRAs must be used with care. For the larger basins that consist of one or more MLRAs, however, the accuracy of the expanded NRI data should be adequate. This assumption needs some further testing and examination.

Components Affecting Water Quality

Assessment of the water quality impacts of agricultural activities requires land-based information and practices that can be linked with hydrologic and toxicological information. The NRI provides data that can be used to estimate sediment movement, and stream loading by inference, but it provides no time-sequenced hydrologic data or direct information on either the amount of fertilizers or pesticides applied to the fields or that transported by sediment or carried in solution.

An assessment of the data and factors most applicable for the assessment of offsite impacts is shown in Table 1. Information from these fields can be used to estimate gross sediment movement, but not deposition. Estimating movement beyond the edge of a field, although possible,

TABLE 1 Data File Fields in 1982 NRI Related to Offsite Impacts

Field	Name
6	Hydrologic unit
10	T factor
12	Degree of erosion
20-22	Land use and cover
24-27	Cropping history
28-30	Conservation practice
32-37	USLE factors
39	Average annual tons of soil 40-41
	Average annual tons of soil loss due to sheet and rill and wind erosion

SOURCE: USDA (1984).

requires the development and use of sediment delivery ratios, a complex and difficult task. By comparing 1977 and 1982 NRI data for a particular area, it may be feasible to determine some trends in sediment loading rates, particularly in areas where there have been significant changes in land use or cropping or tillage practices. However, although the NRI can be used to estimate partial sediment loads, it provides no information on particle size distribution, which is very important for assessing such offsite impacts as fish reproduction. Soils-5 data contain particule size information which can be combined with the NRI.

SOME WATER QUALITY APPLICATIONS OF THE DATA

The NRI data base can be used in conjunction with pollution loading models and more complex water quality models. This section examines some generalized loading model considerations, some specific applications of NRI data with other data bases, and some possible uses of the NRI data with existing water quality models.

Gross Load Estimation

Generalized Procedure

The quantity of a given pollutant passing through a system at any time is the interaction of the process of supply and transport. Several generalized procedures can use NRI data to estimate pollutant loads, which can be useful indicators of the water quality impacts of erosion. One approach is to multiply the Universal Soil Loss Equation (USLE) estimates of soil movement at the watershed or subwatershed level by a sediment delivery ratio and a potency factor, which yields an estimate of the pounds/acre/year of a pollutant (such as sediment, nitrogen, or phosphorus) moving into streams. The potency factor measures the amount of pollutant associated with each unit of sediment (Dean, 1983). The challenge is devising the proper sediment delivery ratio and potency factors for this procedure.

There are several ways to predict or simulate agricultural pollutant loads. These range from simple sediment loading functions to physical processes requiring simulation of chemical reaction, transformation, and dynamic transport.

A pollutant load is defined as a mass of pollutant moving to a receiving water body in a given period of time. If the pollutant is assumed to be linearly correlated with the amount of sediment moving from the watersheds, the pollutant load (or loading function) can be estimated as the product of the amount of sediment delivered to a receiving water body and a potency factor P, which is a factor relating the load of pollutant associated with each unit loading of sediment.

P is very complex and difficult to estimate. As an empirical approximation, the potency factor can be envisioned as the product of the average concentration of a pollutant in the surface layer of the soil, the enrichment ratio of the pollutant of interest, and the ratio of the mean particle density of surface soil to the mean particle density of the eroded sediment (Dean, 1983). The enrichment ratio represents the effect of several processes that cause the ratio of the mass of pollutant to sediment to be higher at stream edge rather than at the source, back at the watershed.

Estimates of the sediment delivery ratios and potency factors for various pollutants have been developed for specific studies, but there are few generalized sets available to apply in large area studies (Gianessi et al., 1981a). Loading rates are related indirectly to the various tillage practices through the interactions of the management components in the USLE calculations. If some figures on changes in use of conservation tillage are available for a given watershed, changes of likely pollutant loadings can be estimated.

The gross loading information from the 1982 NRI provides some information for estimates. However, the data needed for state-of-the-art models to assess offsite effects of agricultural activities are much greater and more complex than those provided by the NRI.

Examples of Loading Models

Haith and Tubbs (1981) developed three loading models to estimate nutrient and pesticide losses from cropland. The models range from simple loading functions to detailed computer simulation models for the soil environment. They share several common attributes. All have a daily time step and are based on the SCS Curve Number Runoff Equations and the USLE. None requires calibration, and each model was tested. Simple planning models or loading functions can provide straightforward means of estimating nonpoint source pollution.

Simple nonpoint source models (loading functions) have several deficiencies. The USLE was not designed to evaluate nonpoint source pollution. It can be used to calculate average annual soil loss, but not loss from single storm events. Although loading functions have been used extensively for this purpose, their ability to provide reasonable estimates of agricultural nonpoint source pollution in large watersheds has not been established. The problem often overlooked in the use of simple loading models is that different categories of potential pollutants are transported in different fashions. For example, dissolved chemicals move with runoff water while most phosphorus and some nitrogen and hydrophobic chemicals are associated with sediment. Loading functions for sediment-associated chemicals should be based on soil loss estimates; dissolved chemicals require runoff-based loading functions.

Specific Applications

Resources for the Future

Resources for the Future (RFF) used the 1977 NRI data base to analyze the relative importance of nonpoint source pollution control options at the national level (Gianessi et al., 1981a,b). This national network model linked point and nonpoint sources of pollution to evaluate agricultural sediment control policies in conjunction with point source controls. The model linked pollution-generating activities in each county to a detailed network of rivers, lakes, and bays. It provides general estimates of the impact of sediment and sediment-bound pollutants on water quality in specified bodies of water. However, it does not evaluate the transport and impact of soluble pollutants.

RFF is updating its national model to incorporate the 1982 NRI data and refining it to include a sediment transport component. This will help in revising estimates for gross sediment and associated pollutants reaching streams.

Economic Research Service

The Economic Research Service of the USDA is using NRI data to estimate offsite benefits associated with soil conservation (Ribaudo, 1984). The NRI information is being used in conjunction with other data sets to relate the levels of pollution associated with erosion parameters to specific impaired water uses. Water quality data from the National Stream Quality Accounting Network (NASQUAN) was used to estimate the ambient water quality levels, total suspended solids, total phosphorus, and total nitrate for each of the 99 watershed units defined by the Water Resource Council as aggregated subareas (ASA). These levels were then compared with standards reflecting impacts on water use.

Pollutant loads in the various watercourses from all sources were estimated using the National Water Discharge Inventory developed by RFF. Sediment discharges from cropland, pastureland, rangeland, and forestland were based on erosion estimates provided by the 1977 NRI. Estimates of streambank, gully, construction site, and other erosions come from other sources. A sediment delivery ratio was calculated for each ASA by RFF and used to estimate the amount of eroded soil reaching waterways. The amounts of total suspended solids, total phosphorus, and total nitrogen in the discharge were estimated using coefficients based on the characteristics of the major soil groups contained in each ASA.

The final step was to compare the pollutant loadings from agriculture with the uses made of the streams in the affected areas and to identify regions where agricultural erosion has significant impact on offsite water uses. Thirty-eight ASAs were identified as having a water quality problem due to agriculture, but only 15 were intensive use regions. The estimates are being updated by incorporating the 1982 NRI and other information into the RFF model.

Linkages with Water Quality Models

Several models have been developed to evaluate the impacts of alternative management strategies on water quality and the influence of specific management practices on the levels of particular pollutants. Some use NRI data, but since most models have gone beyond the gross loading stage, they use the NRI data as one input among several, not as the primary data set. A number of these models may be able to take the erosion estimates as input, but others need only the USLE coefficients in the NRI. Selected examples of models that can use some of the NRI data are described in this section.

A pesticide root zone model (PRZM) being developed by the U.S. Environmental Protection Agency (EPA) at the Athens Environmental Research Laboratory simulates the vertical movement of pesticides in the unsaturated soil within and below the plant root zones (Carsel et al., 1984). The model consists of hydrology and chemical transport components that simulate runoff, erosion, plant uptake, leaching, decay, or surface washoff and volatilization of a pesticide. The hydrology component for calculating runoff and erosion is based on the SCS Curve Number technique and the USLE.

PRZM can be used to estimate frequency distributions of the mass of pesticide leaching from the plant root zone to investigate the risks of pesticide use, particularly pertaining to groundwater pollution. The model uses the Modified Universal Soil Loss Equation (Williams and Berndt, 1977). This modification replaces the R (rainfall erosivity) term with an energy term and allows estimation of the volume of event runoff and peak storm runoff. The model requires all the other USLE factors.

A comprehensive basin-scale simulation model developed to predict water quality arising from both point source and agricultural nonpoint source pollution is the Hydrological Simulation Program in Fortran (HSPF) (Donigian et al., 1983, 1984). The goal of this model is to go beyond the prediction of the quantity and quality of runoff from agricultural lands and to predict instream water quality effects of the best management practices. However, runoff models by themselves are not sufficient to do this, since instream transport and transformations are usually not represented. Using a model of this type requires far more data than the NRI provides. Yet, applying it allows simulation of the movement of pollutants and assessment of the likely impacts of changes in management practices on water quality through time. Models like this are indicative of the state of the art in water quality modeling.

Linkage with the Chemicals, Runoff, and Erosion from Agricultural Management Systems (CREAMS) model developed by USDA is another possible application of the NRI data (Knisel, 1980; Knisel et al., 1983). A major use of CREAMS is evaluation of alternative management practices for control or minimization of runoff of sediment and chemicals. It has three components (hydrology, erosion/sedimentation, and chemistry) and describes the movement of runoff, sediment, and plant nutrients and pesticides from field-sized areas. It is a continuous simulation model that operates efficiently to allow consideration of long-term records (20 years). The model can be used to evaluate the impact of management practices on the yield of sediment and chemical pollutants from field-sized areas at specific sites. It is also being expanded to address questions at the watershed level.

The model's erosion/sedimentation component could use some of the NRI data. It considers the basic processes of soil detachment, transport, and deposition. Detachment is described by a modification of the USLE for a single storm event. The calculation of a rate of interrill detachment and the rate of detachment by rill erosion uses the USLE factors.

Uses of the NRI Data for Offsite Analysis: A Summary

As noted, some inherent constraints in the NRI data base influence its usefulness and applicability for inventorying, monitoring, and appraising offsite erosion damages and adverse impacts on water uses. It provides information on land use and gross erosion estimates, which are useful for loading estimates and models, but it does not have complete data for water quality analysis. Nevertheless, there are several opportunities for its use.

The NRI is a source of coefficients for estimating sediment load from agricultural nonpoint sources, which can be combined with other sources to estimate combined point and nonpoint sediment stream-pollution loads. The applications by RFF (Gianessi et al., 1981a,b) and the water quality assessments by the Economic Research Service (Ribaudo, 1984) are examples. As a screening tool, the NRI data can be used as part of a system to identify areas where sediment and associated pollutant loads in streams really impair stream usage. Comparisons between subbasin characteristics are useful for isolating exceptional situations.

The USLE coefficients in the NRI can be used directly in regional water quality modeling efforts. For example, PRZM needs USLE coefficients to operate, and CREAMS uses USLE factors as part of the input data.

Data from the 1977 and 1982 NRIs has limited use for trend assessment. For particular areas, changes in land use and conservation and tillage practices can be used to estimate changes in gross pollution loads from agriculture, provided sediment delivery ratios and pollution loading coefficients are available. Using the same system, projected changes such as significant shifts to conservation tillage can be analyzed for impact on pollution loads, provided the distinctions are maintained. For example, studies have found a reduction in sediment-transported nitrogen and phosphorus with fluted coulter conservation tillage, and control of both sediment-transported and solution nitrogen and phosphorus with in-row chisel tillage (Langdale and Leonard, 1983).

The NRI can also be part of the data base needed for water quality assessments in specific regions, as done by Ribaudo (1984). It can be coupled with the vast amounts of information already developed in areas of intensive study, such as the Chesapeake Bay (EPA, 1982; Northern Virginia Planning District Commission, 1983). Loading rates for the entire Bay have been estimated. The NRI can be used to estimate loadings from specific basins or subbasins to aid in planning and analysis.

CONSIDERATIONS FOR FUTURE NRIs

Before advocating changes to remedy the NRI's inability to address water quality questions, it is important to remember that the main objective of inventory is to assess the nation's soil and water resources. As such, it was not designed or specifically charged with a responsibility to assess offsite damages. Yet, those charged with designing the next NRI might consider the extent to which it should address offsite questions. Although few data bases are intended to be universal in scope and applicability, with some minor modifications to the NRI, a more complete set of data for water quality analysis could be assembled. Some suggestions for consideration are offered in this section.

A critical missing link in the NRI as far as offsite damage assessment goes is its lack of linkages with hydrologic, toxicological, or meteorological data bases. To get good estimates of sediment load, nitrogen runoff concentration, or phosphorus runoff concentration, the timing of tillage practices and fertilizer applications needs to be tied to meteorological data, particularly rainfall. Time-series data on meterological and hydrologic data are needed for water quality simulation models such as the nonpoint source model (Donigian and Crawford, 1976b, 1977), the agricultural runoff model (Donigian and Crawford, 1976a), HSPF, and CREAMS.

Water quality problems are time-based, but the NRI provides no time-variance loading information. It is impossible to predict water quality accurately from an average annual estimate provided by the USLE. It has validity only for cases where the retention time is 1 year or greater. Linkages are needed on a storm-by-storm basis to rainfall, runoff, soil loss, pollutant concentration, utilization, infiltration, percolation, and movement of soluble pollutants to groundwater. The feasibility of collecting such information in the future needs to be evaluated.

Better links between movement on the field and deposition in the field and streams are also needed. The USLE generates sediment load information, but the sediment delivery ratios and potency factors needed to better assess the amount actually moving to streams are generally inadequate.

Information on management practices and application rates of pesticides and fertilizer would help assess runoff problems as well as the enrichment of sediment. Very little is known about the composition of runoff, either in terms of quantity or quality. Data on fertilizer and pesticide management practices and the properties of the chemicals are particularly important in assessments of whether potential pollutants are moving overland to a stream in solution or bound to soil particles, or whether they are moving down through the soil profile in solution. Research has indicated greater efficiencies in the use of nitrogen in corn and soybeans with no-till than with conventional till (Hoyt et al., 1983). In the Southeast, changes in tillage practices have resulted in greater concentrations of phosphorus and nitrogen, but in reductions of the transported mass (Langdale and Leonard, 1983).

The NRI does not now address the question of ephemeral gully erosion, in which there is considerable interest (see Foster, this volume). Whether it should or could do so needs to be assessed, first to explain the total erosion process better, and second to assess impacts on water quality that fall between rill and gully erosion.

Given the rapid adoption of conservation tillage and no-till, future NRIs need to consider more explicit measurement of these practices (as well as related practices such as fertilizer, insecticide, and herbicide use) and to ensure that they are properly accounted for in the gross erosion calculations. Explicit linkages between the type of tillage and the impact on a specific pollutant and its pathway to stream or groundwater are needed.

There are several excluded sources of erosion needing consideration in efforts to assess the impacts of erosion on water quality. Ephemeral gully erosion, which occurs between rill and gully erosion, is one such source. There is considerable interest in better explaining ephemeral gully erosion as part of the total erosion process; its inclusion in the NRI needs to be assessed. Other erosion not measured in the NRI includes streambank erosion, erosion from federal lands, and erosion from construction sites.

Lastly, assessment of the offsite water quality impacts of soil erosion needs to involve major federal and state agencies with capabilities and responsibilities in the area. Coordination with departments, such as the EPA, is needed if water quality becomes a major emphasis of NRI efforts.

SUMMARY

There are limits on the usefulness of NRI data for assessing the offsite effects of soil erosion. Selected data can be used to estimate pollutant loads, primarily sediment, and thus help identify and inventory potential sources of offsite damages. Data from the 1977 and 1982 NRIs can provide points for assessing trends in changes in the resource use, thus assisting in monitoring factors that influence offsite erosion damages. Data and coefficients from the NRI can be linked with other data bases and water quality models to appraise offsite impacts.

It must be remembered that the NRI was not designed with water quality as its primary focus. Thus, it must be viewed as an important source of information, but useful for addressing water quality questions primarily in conjunction with other models and data. The next NRI could be modified to be more directly applicable to water quality and offsite damage. Issues to consider include improved linkages with hydrologic data bases, linkages with time-based pollutant loads, field-to-stream linkages, linkages to management and tillage practices, the role of ephemeral gully erosion and streambank erosion, and possible coordination with other water quality agencies.

REFERENCES

Bailey, G. W., and R. R. Swank, Jr. 1983. Modeling agricultural nonpoint source pollution: A research perspective. Pp. 29-41 in Proc. of the National Conference on Agricultural Management and Water Quality. Ames: Iowa State University Press.

Bailey, G. W., and T. E. Waddell. 1979. Best management practices for agriculture and silviculture: An integrated overview. Pp. 33-36 in Best Management Practices for Agriculture and Silviculture, Proc. of the 1978 Cornell Agricultural Waste Management Conference. Ann Arbor, Mich.: Ann Arbor Science Publishers.

Carsel, R. F., C. N. Smith, L. A. Mulkey, D. Dean, and P. Jowise. 1984. Users Manual for the Pesticide Root Zone Model (PRZM)--Release 1. (Draft.) Athens, Ga.: U.S. Environmental Protection Agency.

Dean, J. D. 1983. Potency factors and loading functions for predicting agricultural nonpoint source pollution. Pp. 155-177 in Proc. of the National Conference on Agricultural Management and Water Quality. Ames: Iowa State University Press.

Donigian, A. S., and N. H. Crawford. 1976a. Modeling Pesticides and Nutrients on Agricultural Lands. EPA 600/2-76-043. Athens, Ga.: U.S. Environmental Protection Agency.

Donigian, A. S., and N. H. Crawford. 1976b. Modeling Nonpoint Pollution from the Land Surface. EPA 600/3-76-083. Athens, Ga.: U.S. Environmental Protection Agency.

Donigian, A. S., and N. H. Crawford. 1977. Simulation of Nutrient Loadings in Surface Runoff with the NPS Model. EPA 600/3-77-065. Athens, Ga.: U.S. Environmental Protection Agency.

Donigian, A. S., Jr., J. C. Imhoff, and B. Bicknell. 1983. Predicting water quality resulting from agricultural nonpoint source pollution via simulation--HSPF. Pp. 200-249 in Proc. of the National Conference on Agricultural Management and Water Quality. Ames: Iowa State University Press.

Donigian, A. S., Jr., J. C. Imhoff, B. R. Bicknell, and J. L. Kittle, Jr. 1984. Application Guide for Hydrological Simulation Program in Fortran (HSPF). EPA 600/3-84-065. Athens, Ga.: U.S. Environmental Protection Agency.

EPA (Environmental Protection Agency). 1982. Chesapeake Bay Program Technical Studies: A Synthesis. Washington, D.C.: U.S. Environmental Protection Agency.

Gianessi, L. P., H. M. Peskin, and G. K. Young. 1981a. Analysis of national water pollution control policies: 2. Agricultural sediment control. Water Resources Res. 17:803-821.

Gianessi, L. P., H. M. Peskin, and G. K. Young. 1981b. Analysis of national water pollution control policies: 1. A national network model. Water Resources Res. 17:796-801.

Haith, D. A., and L. J. Tubbs. 1981. Operational Methods for Analysis of Agricultural Nonpoint Source Pollution. Ithaca: Cornell Agricultural Experiment Station, State University of New York at Cornell.

Hoyt, G. D., R. L. Todd, R. A. Leonard, L. Asmussen, and E. D. Threadgill. 1983. Effect of tillage on nutrient cycling in Southeastern Coastal Plain agroecosystems. In Nutrient Cycling in Agricultural Ecosystems. Special Publication No 23. Athens: University of Georgia College of Agriculture Experiment Stations.

Knisel, W. G., ed. 1980. CREAMS: A Field Scale Model for Chemicals, Runoff, and Erosion from Agricultural Management Systems. Conservation Research Report No. 26. Washington, D.C.: U.S. Department of Agriculture.

Knisel, W. G., G. R. Foster, and R. A. Leonard. 1983. CREAMS: A system for evaluating best management practices. In Nutrient Cycling in Agricultural Ecosystems. Special Publication No. 23. Athens: University of Georgia College of Agriculture Experiment Stations.

Langdale, G. W., and R. A. Leonard. 1983. Nutrient and sediment losses associated with conventional and reduced-tillage agricultural practices. In Nutrient Cycling in Agricultural Ecosystems. Special Publication No. 23. Athens: University of Georgia College of Agriculture Experiment Stations.

Northern Virginia Planning District Commission. 1983. Chesapeake Bay Basin Model--A Final Report. (Draft.) Annandale: Northern Virginia Planning District Commission.

Ribaudo, M. 1984. Water resource problems. Economic Research Service (ERS) draft materials. Washington, D.C. November 29. Photocopy.

USDA (U.S. Department of Agriculture). 1984. National Resources Inventory--A Guide for Users of 1982 NRI Data Files. (Draft.) Washington, D.C.: Soil Conservation Service.

Williams, J. R., and H. D. Berndt. 1977. Sediment yield reduction based on watershed hydrology. Trans. ASAE 20:1100-1104.

Discussion

Ronald B. Outen

The National Resources Inventory (NRI), as Christensen's paper points out, is just a piece of the puzzle. It does provide some useful information, but alone it is not yet sufficient to make the critical connection between land use practices and water quality.

When water quality problems associated with nonpoint source pollution were first raised years ago, the discussion almost always turned to agriculture, and almost immediately to soil erosion. Yet as Christensen notes, sediment per se is not necessarily the biggest problem in all or perhaps even most of the watersheds, strictly in terms of water quality. Very often the issues most discussed are nutrients from fertilizer and animal waste runoff and, also, pesticides.

Most people make a brief reference to groundwater and then move on to talk about surface water runoff, which is understandable because over the last few years the Clean Water Act has dealt almost exclusively with surface water. There may be a bit of a contradiction in national policy goals in terms of protecting groundwater and surface water. On a given piece of land with an excessive amount of nitrogen, phosphorus, other fertilizer materials, or pesticides, measures to prevent runoff might exacerbate the groundwater problem. More coordination is needed between these programs.

Ultimately, groundwater, like nonpoint source pollution of surface water, must be dealt with on an areawide basis in terms of aquifer recharge areas, at least for those pollutants that tend to be dispersed across the landscape. Moreover, an integrated hydrologic regime that includes both groundwater and surface water must be considered, as well as an integrated land/water network. Furthermore, management practices must be broadly defined.

There soon will be a federal law calling on states to develop implementation programs to apply best management practices in large areas of the country for purposes of water quality. This new nonpoint source management program, which will be added to the Clean Water Act, has received strong support in both houses of Congress.

9

New Cropland in the 1982 NRI: Implications for Resource Policy

Clayton W. Ogg

Comparing the 1982 National Resources Inventory (NRI) with the smaller inventory conducted 5 years earlier shows no startling land use or management changes. Reduced tillage increased, as expected, while a 2.7 percent annual rate of increase in cropland in the mid-1970s was reduced by the pressure of today's lower prices to about 1 percent. Since much knowledge gained from the 1977 NRI was simply reconfirmed in 1982, the more significant insights pertain to new cropland conversions discovered through the recent NRI. This new cropland exemplifies erodibility and other characteristics of soils at the margin of production.

Several U.S. commodity and trade policies influence conversion of these erodible soils and soils with other problems identified in the NRI. For example, commodity price support programs administered by the U.S. Department of Agriculture (USDA) contain an evolving mixture of subsidies, which promote crop production and exports, and land retirement programs, which support prices by reducing crop acreage. Conversion of land to crop uses is affected by these policy choices, as well as by proposed sodbuster and conservation reserve legislation designed to reduce any program-related stimulus to misuse erodible land. The large commodity program outlays relative to conservation cost sharing have focused public attention on farm policy options that affect land conversions.

The 1982 NRI reports both the current use of fields and that of the preceding 3 years. It indicates that the acres cropped in 1982 but not earlier contain a far larger proportion of erodible land than the acres already in production. This paper looks at the new acreage for each major program crop and analyzes erodibility and other physical soil problems, citing earlier estimates of yields and economic factors affecting their use. NRI data are used to consider relationships between soil erosion and current price supports. The paper also demonstrates the economic importance of erodible land at the margin of production and improves on earlier estimates of the erosion-related costs of using this land. It ends with an analysis of policies on land conversions and suggestions on new approaches to policy research using this new and powerful information.

EMERGENCE OF ERODIBLE LAND USE AS A FARM POLICY ISSUE

Conservation has had a role in farm programs for several decades, although not a sharply defined one. Terms like "conservation reserve," "soil bank," and "conservation use acres" suggested some conservation purpose for acreage idled by various price support programs. Yet, until a USDA study in 1984 (USDA, 1984), it was apparently assumed either that farmers would select erodible land for their program acres or that it did not matter enough to be an issue. Cost-sharing programs for applying conservation practices similarly treated erosion as if it were ubiquitous (USDA, 1980).

The 1977 NRI first demonstrated the concentration of erosion problems on a small portion of the nation's cropland. Heimlich and Bills (1984) found only 23 percent of cropland needing additional treatment to reduce sheet and rill erosion. Meanwhile, about a third of the cropland was so disinclined toward sheet and rill erosion as to never need treatment. According to Heimlich and Bills (1984), 8 percent of cropland is so highly erodible that it cannot be continuously cropped without experiencing erosion rates in excess of the 5 tons/acre/year considered allowable. The 1982 NRI gives little reason to alter these figures.

Broad public concern about erosion-related productivity losses and water quality damage evolved into concern focused especially on the small portion of the land that would experience high erosion rates (USDA, 1980). In addition, a dramatic increase in farm exports and in cropland in the 1970s drew attention to erodible land being converted to crop uses. One study suggested that much of the land suited for conversion was going to be more erodible than existing cropland (Cory and Timmons, 1982). The 1982 NRI can now more authoritatively distinguish erosion and other soil problems on land that is actually being planted in crops in the United States.

POLICIES AFFECTING ERODIBLE AND MARGINAL LAND CONVERSION TO CROP USES

Farm policies inevitably influence land conversion to crop uses because they subsidize production and support prices. Farmers Home Administration loans and disaster relief have constituted the major direct subsidies to agriculture. Price supports, meanwhile, consist of required acreage set-asides, paid crop diversions, and grain storage programs.

Since price support programs ultimately rely on acreage reductions to diminish supply and raise prices, their initial impact is to reduce the use of some land that is currently used to raise program crops. As prices go up, however, new land is brought into production, a phenomenon referred to as "slippage." The net program effect in the short run is a reduction of erosion on the idled acres, minus any change from erosion on land brought into cultivation.

Concern that programs may be having a negative overall effect on soil erosion resulted in legislative initiatives to deny program benefits to farmers plowing new fragile or highly erodible soils. Other programs, including a conservation reserve that was tested in the 1984 program, are meant to attract more erodible land into the acreage reduction programs. These worries about the net erosion effects of farm programs underscore the need for greater knowledge about changes in land use in response to program subsidies.

CROPS AND SOIL GROUPINGS IN THE ANALYSIS

Since the 1982 NRI identifies land uses in the 3 years before the sample points were visited, land newly converted to crop uses is easily identified. (The 1982 points that had been sampled in 1977 could also be examined, but definitional changes make that difficult. Also, 1982 is a much larger sample.) There were 10.6 million newly cropped acres in the sample years (mainly 1981 and 1982), mostly in corn, wheat, and soybeans; over 70 percent of these acres were plowed in the previous 2 years as well. To analyze this new cropland, the acreage was placed in six land groups used in a recent study of acreage reduction programs (Ogg et al., 1984).

The land groups were selected to reflect the current policy emphasis on the use of marginal (less productive)

TABLE 1 Land Groups and Yields

Land Capability Class System (LCCS)[a] or Group (Bushels/Acre)		Average Land Corn Yield Erosion Designation
1	I	109
2	IIw, IIs, IIc, IIIw, IIIc, IVw IVs, IVc, and V	67
3	IIe, IIIe, and IVe; RKLS under 50[b]	97
4	IIe and IIIe; RKLS over 50	85
5	IVe; RKLS over 50	79
6	VI, VII, and VIII	37
Weighted average		102

[a]Under the LCCS designations, Roman numerals I-VIII designate severity of the problem for crop uses; subscripts w, s, c, and e indicate whether the problem is due to wetness, stoniness, climate, or erodibility.
[b]An RKLS of 50 implies an erosion rate of about 15 tons/acre/year under average management.

SOURCE: Ogg et al. (1984).

and highly erodible land. Highly erodible land has not been much affected by conservation programs because of the high cost of adequate treatment. Yet, the current cross-compliance and sodbuster regulatory options would tend to affect this highly erodible land more than land with moderate erosion problems.

Soils were grouped by using erosion potential data from the NRI as well as the Land Capability Class System (LCCS) (<u>see</u> Table 1). Soils with few limitations for crop use are in group 1, the most productive land. LCCS data are especially useful for identifying wetness (subclass w), stoniness (s), and climatic limitations (c) for crop uses; soils with these problems were placed in group 2. Productive soils with moderate erosion problems were in group 3, while groups 4 and 5 contain most (71 percent) of the land Heimlich and Bills (1984) described as so highly erodible as to be difficult to treat adequately if in crops other than hay. Group 6 includes a small area of unproductive soils identified by the Soil Conservation Service (SCS) as unsuitable for crop uses, which actually accounts for less than 1 percent of U.S. crop production other than hay.

These six land groups provide an excellent basis for analyzing land conversion to cropland: Earlier estimates of average yields and other economic information that influence land conversion decisions can be drawn on, and four of the groups contain either erodible or lower-yielding soils that are to be expected at the margin of production.

The analysis of acreage reduction programs found that the erodible groups 4 and 5 were not particularly favored by farmers interested in placing their land in current land retirement programs (Ogg et al., 1984). A USDA study (1984) of the 1983 program supported the conclusion that land in acreage reductions is fairly representive of land in crops. However, as mentioned earlier, the land converted to crop use in response to price supports and subsidies was expected to be far more erodible than that idled by programs.

THE NEW CROPLAND

As noted earlier, only 8 percent of U.S. cropland was highly erodible in 1977. However, far more than 8 percent of the new cropland in 1982 in all major crops fell in this category of high erodibility. For example, looking at highly erodible land across all land groups in the 1982 NRI, 30 percent of new corn acres were highly erodible, with 23 percent in relatively productive groups 4 and 5 in Table 2. Average erosion rates for all new cropland were about 1.4 times the national average rate for cropland.

About another 15 percent of cropland in 1977 needed treatment to reduce erosion, but it was classified as fully treatable with conservation practices (Heimlich and Bills, 1984). For all the newly cropped acres in 1982, only 12 percent was in this category of needing treatment, mainly (67 percent) in group 3. The highly erodible groups 4 and 5 thus represent a relatively large portion of the erosion on new cropland.

That the highly erosive land coming into production is in relatively productive groups 4 and 5 emphasizes the

TABLE 2 Distribution of Total and New Cropland by Crop and Land Groups, With Total Acreage by Crop, 1982

Land Group	1982 Crop Acres (Percent)[a]	New Acres in Specified Crops (Percent)[b]					
		Corn	Soybeans	Wheat	Sorghum	Cotton	All Other
1	18	3	3	2	2	15	2
2	27	31	47	27	31	27	37
3	42	32	19	50	37	35	30
4	6	15	18	8	8	0	4
5	2	8	6	4	11	1	2
6	4	11	7	9	12	22	24
Total	99	100	100	100	101	100	99
Total (millions of acres)[c]	404	2.178	1.550	1.948	0.504	0.202	4.167

[a]See Huang et al. (1984) for a more precise definition, since these NRI figures were adjusted using an earlier survey.
[b]"New" acres are those of the specified crops in 1982 that were not in crop production in 1979.
[c]The total acres of corn and soybeans are somewhat suspect due to the early start that was necessary in collecting NRI data. Data collected in 1981 may not have been updated to 1982, and corn and soybeans do rotate in many cases. Thus, some of the reported corn acres may actually have been in soybeans in 1982 and vice versa.

SOURCE: 1982 NRI.

importance of protecting this land when it is not particularly needed to meet food and fiber demands. Research is still measuring the economic significance of both onsite and offsite damage from highly erodible versus less erodible soils, but the highly erodible soils do account for 46 million tons--70 percent--of the new cropland's sheet and rill erosion. This is nearly twice the proportion of erosion from all cropland that is provided by highly erodible land. If market conditions improve, and if the improvement causes sodbusting to increase, highly erodible land groups can be expected to be the source of an increasing share of the sediment entering lakes and streams.

NEW ACRES OF PROGRAM CROPS

New erosion problems are particularly evident for the crops most affected by feedgrain programs. About 2.5 times more of the recently converted crop acres than of all crops in 1977 were highly erodible, but for new corn and soybean acres the equivalent figure was 3.8. Because new corn and soybean acres contain so much erodible land, program decisions for feedgrains could perhaps influence use of substantial acreages of erodible land.

Soils with other physical limitations, such as wetness, figure even more heavily in the land conversion process. The less-productive group 2 accounts for 47 percent of the new soybean land (see Table 2). Still, soils with severe erosion problems expand their share of corn and soybean crop acreage more than other problem soils do.

Since the erosion potential data do not include potential for wind erosion, the erodible land groups in these tables account for less of the erodible wheat acres than was the case for other crops. Even with wind erosion, wheat is one of the crops least susceptible to erosion. A given wheat price support outlay is, therefore, less likely than a feedgrain program to damage nonrenewable resources, even though much of the early support for sodbusting legislation originated in wheat states.

Similarly, groups 4 and 5, with high potential for sheet and rill erosion, are hardly relevant in evaluating program impacts on cotton production because cotton suffers mainly from wind erosion. It is worth noting, however, that 22 percent of new cotton is grown on land group 6. The 22 tons/acre/year combined water and wind erosion rates suggest erosion damage is occurring on the

TABLE 3 Distribution of New Cropland and of Least Profitable Cropland (25 Million Acres) by Land Group

Land Group	New Cropland[a]	Least Profitable Cropland[b]
1	3	3
2	34	37
3	35	37
4	13	15
5	6	5
6	9	3

[a]"New" acres are those in crops in 1982 that were not in crop production in 1979.
[b]See Webb et al. (1984).

SOURCE: 1982 NRI.

new cotton acres. An earlier study (Ogg, 1985) found, in fact, that disaster programs were an incentive to bring 2 million new acres of cotton into the Dust Bowl region during the 1970s.

PRICE SUPPORTS AND PROBLEMS WITH THE NEW CROPLAND

Yet erosion is not the only problem created by land conversions. Table 1 shows a version of the Iowa State Center for Agricultural and Rural Development (CARD) model's estimates of yields on each land group. Along with being erodible, the land groups with the lowest yields in Table 1 account for much of the new cropland. In fact, the distribution of cropland among the six groups parallels the distribution of the least profitable land currently in production (see Table 3), identified in a study using a version of the CARD model (Ogg et. al, 1984). Groups 2 and 6 account for considerably more of newly cropped acreage than of current acreage, while the productive land group 1 accounts for very few of the new cropped acres.

Price support statutes aim to prevent new plowing during crop surpluses because it adds to surpluses and to

TABLE 4 Acres, Erosion, and Per Acre Erosion Rates for New Cropland[a]

	Corn	Soybeans	Wheat	Sorghum	Cotton	All Other	Total
Acres (million)	2.178	1.550	1.948	0.504	0.202	4.167	10.550
Erosion (million tons)	20.595	16.805	15.028	5.502	4.466	25.221	87.617
Erosion/acre (tons/acre)	9.4	10.8	7.7	10.9	21.8	6.1	8.305

[a]The new cropland acres were in crop production in 1982 but not in 1979. As Table 2 notes, some of the figures for crops grown in rotations must be considered approximate. The erosion estimates for each crop are nonetheless considered fairly accurate.

SOURCE: 1982 NRI.

the cost of the supports. Erosion and low profitability greatly add to these social costs of land conversions, as additions to crop surpluses mainly come from either erodible or less profitable land.

Short-Term Effects of Farm Subsidies on Land Conversions and Soil Erosion

Program rules limiting production to base averages thus discourage both new plowing and soil erosion in the short term. These rules also use various subsidies to persuade farmers to place land in set-asides and crop diversions, reducing erosion on the idled acres.

However, the Conservation Use Acres data indicate these programs in 1983 reduced erosion just 0.9 tons/acre/year on wheat acres, 3.7 tons/acre/year on feedgrain acres, and 1.8 tons/acre/year for all crops. Many idle acres were not very erodible or if they were erodible, adequate cover was not established. To add to the erosion and surplus problems, programs that raise prices cause some land to be moved into crops from hay or pasture even on farms not participating in the program. New corn acres eroded at 9.4 tons/acre/year (<u>see</u> Table 4), which is 1.23 times the 1977 rate for all corn acres. These figures shed some light on the erosion due to program-induced land use changes.

A study by Ericksen and Collins (1985) found that every 10 acres idled by farm programs reduced crop production only 5 or 6 acres. The 1983 acreage reduction program idled 31.7 million corn acres from the base acreage and 29.3 million wheat acres, but the reduction in actual cropped acres was only 21.3 million and 20.0 million acres, respectively. Reductions in summer fallow and in soybean acreages make up much of the difference.

Yet, among the 19.7 million acres that shifted into corn and wheat in 1983 were 1.0 million acres that moved out of hay production and probably a larger area that shifted from pasture, although that figure may never be known with precision. The net effect of any program in the short run must include any new erosion that results on the acres converted to crops. According to the new-cropland data, the land shifted from pasture erodes at about 8.3 tons/acre/year (<u>see</u> Table 4) in crops, versus about 1 ton/acre/year when it was in pasture. Assuming the converted hayland has erosion characteristics similar to the new cropland, hay alone would offset 7.3 million tons of erosion of the roughly 102-million-ton program savings for these two crops in 1983.

Including pasture conversions would raise the 7.3 million near-term estimate for new erosion due to program-related land conversions that year. Although lack of reliable data on decreases in pastureland in 1983 makes it difficult to estimate accurately all the short-term erosion and production impacts of the large 1983 acreage reduction, unpublished estimates suggest modest increases in the rate of pastureland conversion that year.

Long-Term Effects of Farm Programs

Base acreages cannot greatly limit new plowing if programs are repeated for several years. Much of the 10.6 million new cropped acres from 1979 to 1982 is therefore somewhat influenced by price supports. Farmers plow up new land partly in response to supported prices or to expand their base acreages, which are the basis for receiving program benefits. Wheat bases, alone, expanded from 45 million acres in 1970 to 91 million acres in 1984 (Ericksen and Collins, 1985).

However, price fluctuations, pressure from banks, and land speculation are also associated with these land conversions (Huszar and Young, 1984). Research has yet to determine how much of the new cropland is plowed to reap program-related benefits versus other reasons.

Smaller program changes, such as the 13 million wheat and feedgrain acres set aside in 1979 and the 9.1 million acres set aside in 1982, do not appear to disrupt the pace of conversion from pasture to cropland. For example, new cropland in wheat and other crops occurred at almost equal increments each year for which data are available. Programs stabilizing or raising farm prices and income obviously influence the rate of land conversion in the long run much more than during the year a program feature is introduced.

Productivity Damage on Land at the Margin of Production

A paper by Doering et al. (1983) for the first time estimated the erosion damage from production of program crops and attempted to relate that damage to farm subsidies. Their study conservatively estimated that an acre-inch of soil is worth about $60 and that productivity losses due to erosion cost 2¢/bushel of wheat and 3¢/bushel of corn. They also noted that new cropland, producing for exports, is surely far more erodible and might suffer several times this rate of damage.

Work is being done to improve these average damage estimates as well as to extend this type of analysis by determining how many additional acres are farmed in response to program subsidies. In the meantime, the marginal erosion damage on new cropland can be suggested much more accurately. Since erosion rates on the corn acres at the margin of production are 1.23 times the rates on all corn acres, the per bushel cost estimate of 3¢ can be raised to 3.7¢/bushel. And the cost estimate for erosion on new wheat acres becomes 2.5¢/bushel.

However, a much larger share of the new cropland erosion comes from highly erodible soils. Thus, there is a need both to improve average damage estimates and to determine separate damage estimates for more erodible soils, which may lose more per ton than soils that are less erodible. Analyzing yield losses from erosion on each of the six land groups described here, for example, would shed light on the economic significance of policy options affecting land conversions. The new NRI data suggest several areas to focus new research and improve on past efforts.

Water Quality Damage from Land Conversions

Although 70 percent of the erosion from new crop acres occurred on highly erodible land, according to 1982 NRI data, this land was dispersed across nearly all the major producing regions, which are listed in Table 5. Farm policy choices affecting conversions of erodible land are therefore of national interest from a water quality standpoint.

Those deciding farm policy can only consider water damage from sediment and related nutrients in the most aggregate terms. Nonetheless, the concentration of sediment and nutrient losses on highly erodible land proved very important to eutrophication problems in a Pennsylvania reservoir (Ogg et al., 1983). And a recent study (Ogg and Pionke, 1986) finds that these Pennsylvania results are due to phosphorus adsorption relationships that have wide applicability.

Phosphorus losses are mainly adsorbed to sediment

TABLE 5 New Cropland Acreage Between 1979 and 1982, By Producing Region

Producing Region	Acres (Millions)[a]
Appalachian	1.205
Corn Belt	2.169
Delta	0.992
Lake	0.614
Mountain	1.381
Northeast	0.382
Northern Plains	1.548
Pacific	0.699
Southeast	0.908
Southern Plains	1.178
Total	11.076

[a]Includes some land that shifted in and subsequently out of crop uses.

SOURCE: 1982 NRI.

particles in the case of the highly erodible fields, while phosphorus moves off less erodible fields in its dissolved form. These dissolved phosphorus losses from less erodible soils are not much affected as acres are converted to crops. In addition, pasture or forest uses lower the soil's fertility status. The 70 percent figure for sediment originating from the highly erosive new cropland acres considerably understates, then, the share of phosphorus damage associated with conversions of highly erodible land.

PREVENTING SODBUSTING DURING PRICE SUPPORTS

Since land conversion appears to be a long-term investment decision, provisions that prevent farmers from expanding crop acreages during a particular price support year may not address this process. Meanwhile, the high proportion of the new cropland in the erodible and relatively productive land groups 4 and 5 suggests that sodbuster legislation that includes productive land would help prevent resource problems and new production during price support programs.

The immediate social costs of preventing new production through sanctions on sodbusting are apparently lower than idling acres already in production. As Table 3 indicates, the new cropland is among the least productive now in production, containing virtually no acreage in the land group having the highest yields and substantial acreage in groups with yields one-third to two-thirds below average. The average corn yield on new cropland (estimated by weighting acres in Table 4 by the yields in Table 1) is only 78 bushels--about 23 percent below the average on all cropland. Also, much of the new cropland came from regions like the Southeast (see Table 5), where yields in each land group are below the national average yield for that group. Thus, 23 percent is a conservative estimate of this yield difference.

The plowing costs and other expenses associated with plowing less productive new land therefore adds to the social cost of farm programs to the extent that production shifts from idled program acres to the new cropland acres. According to Watts et al. (1983) it costs at least $134/acre to convert rangeland to crops in Montana. Sodbusting costs less than that in some states, but it represents a substantial expenditure. Erosion, water quality damage, low profitability, plowing costs, and additions to crop surpluses all point to the importance of rules that reduce incentives for sodbusting.

OTHER OPTIONS THAT REDUCE CONVERSION OF MARGINAL OR ERODIBLE SOILS TO CROP USES

In the Appalachian, Delta, Northeast, and Southeast regions, commodity program participation is lower than in the areas that produce wheat and feedgrains. However, credit subsidies are concentrated in the Southeast, so credit policy can be used to discourage conversions of erodible land in these regions (Ogg, 1985). In the rest of the United States, commodity program benefits carry much more weight.

Although sodbuster legislation and conservation reserves affect these land conversions most directly, a bid system used to retire land in 1983 is also relevant. The bid system primarily reduces the farm program outlays by encouraging rental bids from farmers based on the earnings of each piece of land. Bidding would thus reduce the windfalls to owners of less profitable, recently converted cropland (see Table 3). These windfalls are often very large when acreage reduction programs include 50 to 90 percent of western wheat farmers, as they presently do. Deficiency payments and crop diversion payments now offer all program participants for each bushel of their production capacity the amount that the more profitable farmers in the country are earning. Those who produce at low or negative profits capture the largest windfall.

Although a number of factors listed here contribute to the land conversions (Huszar and Young, 1984), one financial incentive for farmers, bankers, and speculators is ultimately to add to the current or future owner's base acreages and capture the windfall described above. The bid system reduces this windfall, encouraging each farmer to idle land for what it is worth. Short of freezing the base acreage, there may be no more effective way to reduce the incentive to expand base acreages.

Research needs, then, to specifically address program incentives, such as the bid system, to prevent base acreages from expanding onto less productive soils. The NRI and some of the modeling tools discussed in this volume will play a role. Such analyses are as relevant to commodity policy as they are to resource concerns. The 1982 set-aside was barely able to offset land conversions of the previous 3 years.

Fundamental research regarding onsite and offsite erosion damage on new cropland is also needed, as others have indicated. (See papers by Benbrook et al., 1984; Christensen, this volume; and Walker and Young, this volume.)

CONCLUSIONS

Land at the margin of production, recently converted to crop uses, experiences lower yields and more erosion than land currently in crops. Much of the new cropland is from the most erodible land groups. Although it also has lower yields on the average, land with wetness and other problems account for the lowest yields, as the erodible land among the new acres comes from fairly productive land groups.

When crop surpluses exist, farm programs attempt to discourage participants from planting new acres in the program crops because the additional cropland undermines the ability to control surpluses. The analysis in this paper suggests that the new plowing, which takes place partly in response to price supports, has the additional undesirable effect of moving the country toward a less efficient agriculture and adding to the nation's water pollution problems.

The immediate economic inefficiencies of plowing new cropland include conversion costs and a shift toward less productive land than that currently in use. This short-term social cost is in addition to erosion damage, which exceeds not only similar damages on land currently in production but also erosion on acres idled by farm programs (USDA, 1984).

These findings suggest that land conversions due to farm policy choices need further study. The erosion and yield analysis of land conversion needs to be expanded along the lines of a recently completed analysis of acreage reduction programs. For specific sodbuster provisions or other policy choices affecting one or more major crops, the physical and economic information is now available to anticipate yield, costs, and erosion impacts for 105 local producing areas in the United States.

REFERENCES

Benbrook, C. M., P. Crosson, and C. W. Ogg. 1984. Resource dimensions of agricultural policy. Proc. Agricultural and Food Policy Conference, Berkeley, Calif. Berkeley: Giannini Foundation.

Cory, D., and J. F. Timmons. 1982. Responsiveness of soil erosion losses in the Corn Belt to increased demands for agricultural products. J. Soil Water Conserv. 33:221-226.

Doering, O., A. Schmitz, and J. Miranowski. 1983. Farm costs and exports. Pp. 117-127 in Increasing Understanding of Public Problems and Policies. Oak Brook, Ill.: The Farm Foundation.

Erickson, M. H., and K. Collins. 1985. Effectiveness of acreage reduction programs. Pp. 166-185 in Agriculture-Food Policy Review, No. 530. Washington, D.C.: USDA Economic Research Service.

Heimlich, R. E., and N. L. Bills. 1984. An improved soil erosion classification for conservation policy. J. Soil Water Conserv. 39:261-266.

Huang, W., B. English, S. Webb, and C. Ogg. 1984. A national linear programming model and data base for analysis of soil erosion. Unpublished.

Huszar, P. C., and J. E. Young. 1984. Why the Great Colorado Plowout? J. Soil Water Conserv. 39:232-234.
Colorado Plowout? J. Soil Water Conserv. 39:232-234.

Ogg, C. W. 1985. Cross-compliance and fragile croplands. In Farm Level Impacts of Adopting Cross-Compliance Programs: Policy Implications. Proc. American Agricultural Economics Association Annual Meeting, Purdue University, West Lafayette, Ind. Columbia: Department of Agricultural Economics, University of Missouri.

Ogg, C. W., and H. B. Pionke. 1986. Water quality and farm policy initiatives. J. Soil Water Conserv. March-April 1986. In press.

Ogg, C. W., H. B. Pionke, and R. E. Heimlich. 1983. A linear programing economic analysis of lake quality improvement using phosphorus buffer curves. Water Resour. Res. 19:21-31.

Ogg, C. W., S. Webb, and W. Huang. 1984. Economic analysis of acreage reduction alternatives including a soil conservation reserve and competitive bids. J. Soil Water Conserv. 39:379-383.

USDA (U.S. Department of Agriculture). 1980. National Summary Evaluation of the Agricultural Conservation Program: Phase I. Washington, D.C.: USDA Agricultural Stabilization and Conservation Service.

USDA (U.S. Department of Agriculture). 1984. Conservation Use Acres Study. Washington, D.C.: USDA.

Webb, S., C. W. Ogg, and W. Huang. 1984. Economic analysis of acreage reduction strategies. Paper presented at the American Agricultural Economics Association Meeting, Ithaca, N.Y., August 6, 1984.

Watts, M. J., L. D. Bender, and J. B. Johnson. 1983. Economic incentives for converting rangeland to cropland. Bull. 1302. Boseman: Cooperative Extension Service, Montana State University.

Discussion

Wesley D. Seitz

Ogg's paper focuses attention on the quality of the land being added to the production base compared with the quality of that already in crops. Quality is measured in terms of productivity and susceptibility to erosion. Evidence suggests that it may be appropriate to develop policy initiatives to control, or at least influence, land conversion decisions because significant acreages of less productive and more erosion-prone land are being brought into production.

An attempt to develop federal "sodbuster" legislation is an indication that the policy process is already sensitive to one manifestation of this problem. As Ogg indicates, the problem is more pervasive; therefore, additional legislative initiatives might be beneficial over time.

A fine tuning of this analysis would be helpful, although it is not clear whether the data are available in the NRI to do this. The average productivity of land being added is lower and the average erosion rate is higher than that of land in the productive base. However, this is not adequate to suggest that a policy of prohibiting all land conversion should be adopted. It is reasonable to expect that some land coming into production has higher productivity and lower erosion rates than some land in the base. If the objective is to improve the quality of land in production, the thrust of policy development should be to slow or stop the conversion of poor, erodible land to row-crop uses, while allowing land with low erodible potential to move into the production base. Policies that would reduce the intensity of production on erodible land currently in the production base would also be consistent with this objective.

A bid system for taking land out of production in any given year has substantial appeal, although there may be regional equity consequences that would make it difficult to implement. It is possible that in some areas nearly all land would be bid out of production, with obvious adverse consequences for input and processing companies. Other areas would be unaffected.

A closely related alternative, purchasing easements for crop production rights on erodible land, has been previously mentioned. In this case, the operator may use the erodible land to produce crops that do not require annual tillage. In this sense, it is equivalent to purchasing mineral rights. Several aspects of this policy are worthy of mention. First, by precluding intensive cropping options, easement purchase would to some degree address the current problems of excessive production. Second, it would be possible to allow land to return to production. Third, because the most erodible land is often also the least productive and the least profitable, it should be possible to attract large acreages of land out of production at relatively low cost per acre. This is not always the case, however, and different policy approaches would be necessary in those areas.

Given world economic conditions, it seems that in the intermediate term, deflated crop prices will drift downward or hold constant. This trend is suggested by the current emphasis in the policymaking arena on moving toward a free market, a foreign trade orientation in establishing crop prices. Some land may therefore be taken out of production by operators for the obverse of the reasons suggested by Ogg that it be brought into production.

In the current political climate, it is not reasonable to expect increased funds for conservation programs in the next fiscal year. When policy formulators make the difficult decisions concerning the expenditure of limited dollars, they are going to respond to the current farm financial crisis. It is real. Many farmers will be or are going bankrupt, and they are going to attract the politicians' attention. Actions designed to address this problem, either in the short term or the long term, are likely to take priority over soil erosion problems in the development of the 1985 farm bill.

However, over the longer run, the funding for erosion control may receive a higher priority. It is reasonable to expect that offsite damages from erosion, such as sedimentation, are going to substantially overshadow the productivity damages associated with erosion over the next 5 to 10 years. Sedimentation of reservoirs, drainage ditches, and harbors is going to be a bigger problem than the onsite damages used in determining T values, the ubiquitous soil loss tolerance limits. It may be worthwhile to begin now to develop a tolerance limit based on the offsite damages associated with erosion, perhaps designated τ. (The wave-like shape is an appropriate symbolism.)

If water quality-based soil loss limits, τ, were established, in many cases they would be more restrictive than the productivity-based limits, T. If the τ limits are to be implemented, it will be incumbent on the public sector to do so.

It seems that tolerance limits were set a number of years ago, based upon general impressions of erosion rates appropriate to allow sustained productivity. Research now under way may allow the development of more accurate assessments of a tolerance limit reflecting the impacts of productivity damages. Refining such a set of erosion limits or targets would be extremely helpful in the development of more robust models at the farm, watershed, state, and national levels. These models would facilitate the analysis of policy alternatives that might provide the means of developing significantly more efficient, production-oriented policy responses.

Discussion

Marion Clawson

Ogg's paper raises four questions about land that has recently gone--or will go--into production. First, why was this land not cropped in 1979? Was it a concern over erodibility, or was it something else? What was the circumstance that led the farmer not to crop it at that time? This is one place where researchers might look for factors other than erosion.

Second, why was this land brought into cultivation in these 3 years? The paper implies--rightly, no doubt--that government programs had something to do with it. But on the other hand, just a very few years earlier, farmers were enjoying some of the most favorable crop prices ever. Why were these particular acres not plowed up in the early 1970s instead of the late 1970s? One answer might be that the farmers have more confidence in their political power to get government subsidies than they have faith in the competitive market to provide them with favorable prices.

Third, is there a lot of additional land not now in crops that is near the margin of development? More explicitly, what are the projections of the rate at which additional land will come into development over the next few years or longer? As pointed out, crop acreage reduction programs have nearly offset new land development. In other words, U.S. farmers have been running hard to stay in the same place; it seems they are going to have to continue to run hard to stay in the same place.

Lastly, what changes in farm organization and farm management are implicit in these recent land developments? It is assumed that all the land developed between 1979 and 1982 was already on farms, and must have precipitated shifts in farm organization and farm management.

These questions, although primarily economic ones, might have considerable implications for soil erosion in the future. There has been a shift away from livestock and toward crops. Will there be a shift back?

10
A Midwestern Perspective on Targeting Conservation Programs to Protect Soil Productivity

C. Ford Runge, William E. Larson, and *Glaucio Roloff*

Soil erosion has been identified as an important potential threat to long-term agricultural productivity in the grain-growing regions of the Midwest. To date, however, much of the evidence supporting this view has been fragmentary or impressionistic. This study uses recent data developed as part of the National Resources Inventory (NRI) by the Soil Conservation Service (SCS) of the U.S. Department of Agriculture (USDA) to assess the potential onsite long-term productivity losses due to soil erosion in six Major Land Resource Areas (MLRAs) of the region.

The essential purpose of this study is to demonstrate how the NRI can help implement policies that increase the efficiency of soil and water conservation by targeting those sections of the land mass most susceptible to damage from erosion. The NRI data, when combined with productivity measures developed by Larson et al. (1983), allow policymakers to go beyond simple measures of potential soil loss, such as topography, to investigate specific soil types that are highly susceptible to productivity declines. By carefully specifying the differential impact of water erosion on these soils, a clearer picture of potential productivity losses can be developed. This study, one of the first of its type, represents an initial effort in what is hoped will be an increasingly refined study of erosion impacts. It must be emphasized that this is preliminary and should not be interpreted as a sufficient basis for policy prescription.

METHODOLOGY

The NRI is a USDA nationwide survey of private agricultural lands that contains data on approximately 22 parameters affecting potential agricultural productivity. Included are both physical characteristics of the land and water resource base and the impact of different agronomic practices on soil erosion. The 1982 NRI expands a 1977 data base, updated to encompass a variety of measures that would allow estimates of erosion potential on different land classes. In addition to the NRI data, this paper utilized the Soils-5 data base established by the SCS (USDA, 1983), which contains soil descriptions, ranges of soil and chemical properties, crop yields, and land capabilities and limitations for U.S. soils.

Together, these data allowed the development of three scenarios that simulate the impact of three stylized programs of soil conservation. In this paper, attention is restricted to the onsite effects of water erosion upon soil productivity for land in row crops (corn and soybeans) in six MLRAS of the Midwest (see Figure 1). Three of these--MLRAs 105, 109, and 113--are highly susceptible to erosion and have soils that may suffer large productivity declines if erosion occurs. The other three--MLRAs 103, 108, and 115--are comparatively less susceptible to erosion or are less likely to suffer large productivity declines if erosion occurs (Pierce et al., 1984). The purpose of this exercise is to compare the impact of alternative soil conservation targeting policies on row-crop production in these areas, using concepts recently developed by Pierce, Larson, and others (Larson et al., 1983; Pierce et al., 1983).

Soil and water conservation programs may be targeted according to myriad criteria, each of which may carry implications for the mix of crops grown and the future productivity of the targeted and nontargeted areas. It is therefore essential to define both the criteria employed and the measures used to estimate their effects. In this paper, three basic scenarios and two measures are used to simulate alternative policies. In each scenario, the two measures reported are: the acres planted to corn and soybeans in each MLRA, and the 100-year impact of this pattern on soil productivity. The first scenario is a baseline estimate of the long-term effects of current erosion rates and the long-term productivity of soils in the six MLRAs if there are no changes in soil and water conservation programs or practices recorded by the 1982 NRI.

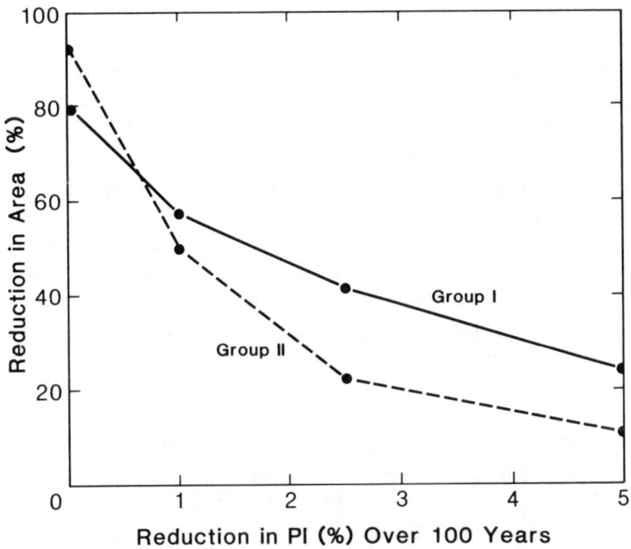

FIGURE 1 Major Land Resource Areas studied.

The second scenario estimates the acreage in each MLRA that must be removed from row crops if a particular tolerance to soil erosion (T value) based on the Universal Soil Loss Equation (USLE) were chosen as a basis for policy. In this case, all land in row crops in the MLRAs under study with an erosion rate greater than the local soil loss tolerance limit (T) for the particular soil series would be put into forage. Given this, the T-value criterion leads to reductions in acreage planted to row crops. The magnitude of these reductions, together with the soil-productivity impact of the shift into forage, is estimated using a 100-year horizon. It is assumed, as in the baseline scenario, that conservation practices continue at current levels. In this scenario, however, these practices encompass those used on the row crops that remain in production, and those used on the land given over to forage (assumed to be the practices considered proper for an established stand with about 80 percent ground cover). This simple scenario can, of course, be modified to include improved conservation practices or other factors such as the relative impact of planting to other crops rather than shifts to forage. Here, however, the estimate is simply of the impacts that might result from such shifts into forage, assuming

T values are the targeting criterion. Clearly, more complex patterns of land use would actually occur.

The third scenario applied to targeting is more refined and involves use of the Soils-5 data base in connection with the 1982 NRI. In this case, it is assumed that some of the land taken out of row crops and put into forage under the second scenario can be returned to row-crop production because the soil type is not highly vulnerable to losses in productivity. The potential productivity losses of soils in each MLRA, as calculated from the vulnerability (V value) of various soil types to erosion, is used as the basis for this targeting criterion. The total acres remaining in row crops are then reported. Because this vulnerability is a measure of potential loss in soil productivity due to erosion, results are reported for four levels of productivity over the 100-year horizon to test the sensitivity of the analysis to these levels. The four cases tested are based on estimates in which productivity declines at a rate 5.0, 2.5, 1.0, or 0.01 percent of its present level over a century.

For ease of reference, the three scenarios employed in this study may be thought of as a no-change baseline, targeted programs based on T values alone, and programs based on new information concerning the vulnerability or V values of various soil types. The three simulations thus demonstrate the way in which alternative targeting policies can be tested using the NRI and Soils-5 data bases. (For more precise definitions and details of the methodology, see the Appendix.)

RESULTS

Baseline Scenario

The first group (hereafter, Group I) of MLRAs (105, 109, and 113) represents areas highly susceptible to erosion and soils that may suffer large productivity declines if erosion occurs. The second group (hereafter, Group II) of MLRAS (103, 108, and 115) is comparatively (though not uniformly) less vulnerable. These regional characteristics are presented in Table 1, the first column of which shows the comparative erosion potential of the six MLRAs in tons/acre/year. This erosion potential, which varies from 11.2 in MLRA 103 to 71.6 in MLRA 105, does not necessarily correspond to levels of

TABLE 1 Characteristics (Weighted Averages) of the Area in Corn and Soybeans, MLRAs 103, 105, 108, 109, 113, and 115, for 1982

MLRA	Erosion Potential	CP Value[a]	Erosion Rate (Tons/Acre/Year) Actual	Erosion Rate (Tons/Acre/Year) Tolerable[b]	Productivity Index	Vulnerability Value
Group I						
105	71.6	0.19	11.4	4.7	0.84	0.23
109	45.1	0.36	15.2	3.9	0.79	0.17
113	25.4	0.33	8.0	3.4	0.72	0.21
Group II						
103	11.2	0.38	4.1	4.9	0.88	0.27
108	28.9	0.36	8.8	4.9	0.91	0.16
115	32.0	0.35	10.2	4.6	0.83	0.14

[a]Cover and management, and support practice factors.
[b]Soil loss tolerance limit (T value).

SOURCE: 1982 NRI and USDA (1983).

conservation practice (CP value) in the MLRAs investigated. In MLRA 105, however, the highest level of erosion potential (71.6) is combined with the lowest relative CP value (0.19), which signals the most intensive level of conservation practice. In MLRA 103, the lowest potential erosion level (11.2) is matched by the highest relative CP value (0.38), indicating the least intensive conservation. Still, these averages tend to obscure cases of potentially poor conservation practices on highly erosive soils within a given MLRA.

This result is suggested by data in Table 1 showing actual and tolerable erosion rates. Tolerable as used here is defined as the soil loss tolerance limit (T value). Intensive levels of conservation practice on MLRAs with high erosion potential, such as 105, still resulted in actual erosion rates of 11.4 tons/acre/year in 1982, more than twice the tolerable level. In MLRA 109, on the other hand, a CP value of 0.36, indicating relatively nonintensive conservation practices, was associated with very high erosion rates of 15.2 tons/acre/year, about four times the tolerable level of 3.9. This suggests that within the MLRAs, as well as among them, more accurate targeting of conservation practices is required. In only one--MLRA 103--was the actual erosion rate in 1982 less than that considered tolerable.

These findings also suggest the need for a sharper analytical tool than is provided by tolerance levels alone. Ideally, such a tool should be able to distinguish, both between MLRAs and within them, which soils are most susceptible to productivity losses due to erosion. This is the purpose of the productivity index (PI) and vulnerability values reported in Table 1. These values indicate, based on specific depths and types of soils in the Soils-5 data base, which areas are a suitable environment for continued crop productivity. The PI of these MLRAs considers the sufficiency of available water capacity for each soil, the sufficiency of soil bulk density, the soil's acidity, and the depth of soil horizons in the zone of plant rooting. The V value is simply the slope of the productvity index/soil removal curve, which plots the loss in soil productivity resulting from incremental reductions in soil depth. (For purposes of this analysis, a linear approximation to this relationship is used. In general, this is quite accurate, although some soils manifest nonlinear productivity losses, leading V to change as successively more soil is removed. V values are reported in this paper as absolute numbers.)

Table 1 shows that, in general, low average levels of erosion potential are generally associated with high average levels of soil productivity, and vice versa. However, the relationship between soil productivity and erosion potential is far from straightforward. Productive soils may or may not be on highly erosive lands, suggesting the need for additional information if policy is to be correctly formulated. On shallower soils, for example, damage provides its own form of conservation incentive, while on deep soils greater rates of erosion may be economically rational (Walker, 1982). In cases where highly productive soils are found on erosive lands, it may nonetheless be appropriate from an economic perspective to continue farming them, and to focus the lion's share of conservation practices there. This requires a measure of soil vulnerability to losses in productivity (the last column of Table 1).

Consider the situation in MLRA 105, in which the highest average level of erosion potential is paired with the most intensive average conservation practices. The productivity index is 0.84, greater than in both MLRA 109 and MLRA 113, where conservation practices are nearly half as intensive on average and where the average productivity indices are lower. The relative potential productivity of the soils in MLRA 105 suggests strong reasons why it should continue to be targeted for improved conservation practices. This argument is reinforced by the overall vulnerability of its soils (0.23), which is greater than in either MLRA 109 (0.17) or MLRA 113 (0.21).

A second example is MLRA 103, which pairs the lowest average level of erosion potential with the least intensive average levels of conservation practices. This relationship appears to indicate no excessive soil erosion losses. However, the high productivity index of the MLRA (0.88) combined with its relatively vulnerable soils (0.27) suggests that the observed levels of conservation practices may not be responsive to the vulnerability to productivity losses due to erosion.

The efficiency losses resulting from a failure to target the soils most vulnerable to productivity declines have been estimated by Ervin et al. (1984). They concluded that "T values may not be appropriate compliance criteria across different soils," and that attempts to target conservation incentives based on T values alone may "give greatest incentives to control erosion on lands for which the long-run social benefits are negative or smaller than for more erosive lands" (Ervin et al., 1984, pp. 277-278). The primary requirement for policy is, therefore, a more accurate targeting criterion based on potential productivity losses.

An important point emerging from this analysis concerns the loss of information resulting from MLRA averages. Although such averages are the basis of this paper, the NRI and Soils-5 data bases allow the development of much more disaggregated simulations analogous to those presented here, which can then be used as a basis for more localized targeting policies.

The next phase of this analysis concerns the comparative impacts of different types of targeting criteria on total acreage planted to row crops (see Table 2). A baseline scenario is given first, in which the acreage in corn and soybeans in each MLRA in 1982 is reported. Based on the assumed continuance of the level of conservation practices reported in Table 1 above, the estimated impacts of these practices on soil productivity over 100 years are then calculated. The highest levels of soil productivity losses are observed in Group I, while the lowest levels would occur in Group II, (hence the rationale for the groupings).

T-Value Scenario

Scenario 2 estimates the impact of using T values as a basis for shifting lands out of row crops and into forage. Wherever the actual erosion rate exceeds T in a given sampling location, these acres are assumed to be taken out of row crops and shifted into forage. The acreage thus removed is subtracted from the baseline acreage. The acreage remaining in row crops is reported, together with the percentage reduction in this acreage and the change in soil productivity resulting from this shift in land use, again assuming no change in conservation practices. The result, as shown in Table 2, would be considerable reductions in 1982 row-crop acreage, ranging from a minimum of 22.0 percent in MLRA 103 to a maximum of 65.6 percent in MLRA 109. As expected, Group II MLRAs would show lower percentage reductions than Group I, although even these are hardly modest. In both groups, radical changes in land use are implied by the T-value targeting criteria.

TABLE 2 Acreage in Corn and Soybeans and Weighted-Averages of the Reduction Over 100 Years in Productivity Index (PI) and Acreage Under Each Scenario, MLRAs 103, 105, 108, 109, 113, and 115

MLRA	Scenario 1 (Baseline)		Scenario 2 (Use of T)			Scenario 3 (Use of V)							
						ΔPI = 5%		ΔPI = 2.5%		ΔPI = 1%		ΔPI = .01%	
	Acres (100)	ΔPI[a] (Percent)	Acres (100)	ΔAc.[b] (Percent)	ΔPI[a] (Percent)	Acres (100)	ΔAc.[b] (Percent)	Acres (100)	ΔAc.[b] (Percent)	Acres (100)	ΔAc.[b] (Percent)	Acres (100)	ΔAc.[b] (Percent)
Group I													
105	33,646	5.6	15,930	52.7	1.0	27,273	18.9	23,958	28.8	18,220	45.8	6,079	81.9
109	31,488	6.9	10,839	65.6	0.8	22,097	29.8	16,395	47.9	10,390	67.0	4,496	85.6
113	27,008	4.4	10,295	61.9	0.7	20,632	23.6	14,166	47.5	10,928	59.5	7,908	70.7
Group II													
103	130,178	1.9	101,485	22.0	0.8	119,673	8.1	101,948	21.7	56,775	56.4	3,035	97.7
108	142,643	2.4	78,608	44.8	0.5	128,580	9.9	112,347	21.2	72,622	49.1	4,909	96.6
115	63,545	3.6	31,990	49.6	0.7	54,176	14.7	47,915	24.6	34,610	45.5	10,933	82.8

[a]Reduction in PI over 100 years.
[b]Reduction in acreage compared with Scenario 1.

SOURCE: 1982 NRI; USDA (1983).

Unsurprisingly, the consequence of these shifts in land use is to reduce substantially the loss in productivity shown in the baseline. In MLRA 109, for example, an estimated 100-year loss in productivity of 6.9 percent would be reduced to 0.8 percent. Comparable reductions occur in the other MLRAS, with the effects most pronounced in Group I. The opportunity cost of these reductions, in terms of acres of row crops foregone, appears to be very large, however, and would lead to major shifts in agricultural production away from these crops.

V-Value Scenario

Scenario 3 uses the information contained in the NRI and Soils-5 data files to target more accurately those soils highly vulnerable to productivity losses due to erosion. The V values listed in Table 1 were the basic criterion used to determine whether lands should be shifted out of row-crop production. This use of V values requires an explicit determination of the rate of reduction in the productivity index over the relevant time horizon. Four levels of reduction are used, both to test the sensitivity of the criterion to judgments concerning productivity declines and to indicate the importance of making such judgments explicit. These judgments must reflect both private and social values concerning the appropriate rate of depletion of soil resources--judgments that are ultimately normative.

For some perspective on the rates chosen (5.0, 2.5, 1.0, and 0.01 percent over 100 years), they may be compared with either Scenario 1 or Scenario 2. A rate of 5.0 percent over 100 years, for example, is comparable to the Group I average of 5.6 percent in Scenario 1, resulting from a policy of no changes in row-crop production or conservation practices for these MLRAs. A rate of 2.5 percent is half this level of depletion, and it is comparable to the Group II average of 2.6 percent in Scenario 1, again implying a policy and set of conservation practices (for this group of MLRAs) essentially the same as at present.

Scenario 2, as noted above, implies much lower rates of soil productivity depletion in return for major land use shifts. Even in Scenario 2, however, the lowest rate of decline is 0.5 (in MLRA 108). The choice of 1.0 as a rate of productivity decline would correspond to MLRA 105 under Scenario 2. The choice of 0.01 percent as an acceptable change in productivity index reflects the notion that almost no depletion is acceptable (a value of zero could not be used for computation).

The resulting estimates in Scenario 3 are instructive. Where 5.0 percent is used as an acceptable 100-year rate of soil productivity loss, a major share of the acreage taken out of row-crop production in Scenario 2 would be returned in Scenario 3. In Group I, MLRA row-crop acreage reduction would average 24.1 percent of 1982 acreage. This compares quite favorably with the average reduction in Scenario 2 of 60.1 percent--roughly 2.5 times as much acreage shifted out of row crops. In Group II, the reduction would average only 11.0 percent of 1982 acreage. In contrast, use of the T-value criterion in Scenario 2 led to an average reduction of 38.8 percent, slightly more than 3.5 times as much acreage. When account is taken of the fact that 1982 was a high-production, nearly record-setting year for these row crops, the actual acreage reductions necessary to achieve 5 percent losses in soil productivity over 100 years if a V-value criterion is used would appear to be even less than suggested by these estimates.

If a stricter soil-productivity-loss criterion of 2.5 percent is applied, the acreage that could be returned to row-crop production in Scenario 3 compared with Scenario 2 drops. In Group I, the reductions implied would average 41.4 percent. Although substantial, this is still considerably below the Group I average reduction of Scenario 2 (60.1 percent). In Group II, the acreage reduction would average 22.5 percent, again substantially less than the drop that would occur under Scenario 2 (38.8 percent).

Results for a 1.0 percent rate of soil depletion led in general to reductions of the same order of magnitude as the T-value criterion used in Scenario 2. The only exception is in MLRA 103, where the relative vulnerability of soils would lead to greater reductions in acreage. This suggests the greater precision of the vulnerability measure. Overall, the implicit rate of depletion resulting from use of T values is approximately 1.0 percent over 100 years.

The strictest assumption--of only a 0.01 percent reduction in soil productivity over 100 years--gives some indication of the magnitude of land use shifts that would be necessary to pursue essentially "steady-state" policies with respect to soil loss. In the case of Group I, 79.4

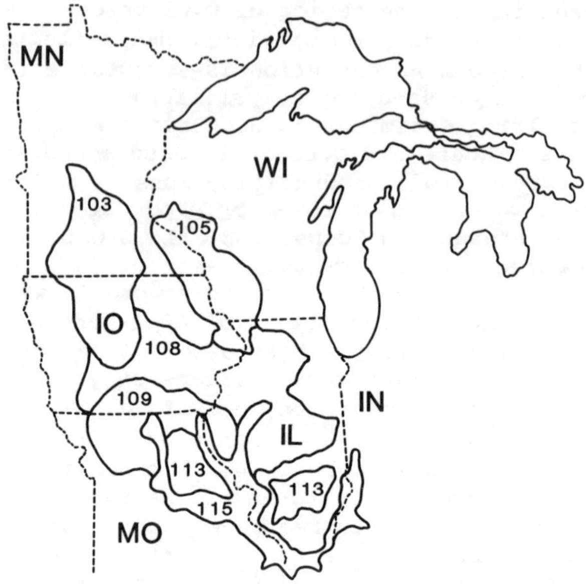

FIGURE 2 Average reduction in (corn and soybean) acreages for a given reduction in productivity index over 100 years, for MLRAs 105, 109, 113 (Group I) and 103, 108, 115 (Group II).

percent of all row-crop acres in 1982 would have to be shifted to forage on average. In Group II, an average of 92.4 percent of all 1982 row-crop acres would be pulled from production and put into forage. In short, pursuit of a "steady-state" level of soil productivity implies the elimination from row-crop production of the vast majority of the acres in those MLRAs.

The overall relationship between reductions in acreage planted and reductions in the percentage of productivity lost in Scenario 3 is shown in Figure 2. As less reduction in productivity is allowed over 100 years, proportionately larger shares of the acreage planted to row crops is removed from Groups I and II. As the figure shows, the distribution of vulnerability differs, and the acreage taken out of production rises at an increasing rate as the requirements for maintained productivity converge to the "steady state."

In all, these results suggest that substantially fewer acres could be shifted from row crops to forage if targeting policies for soil and water conservation were based on soil vulnerability to productivity losses rather

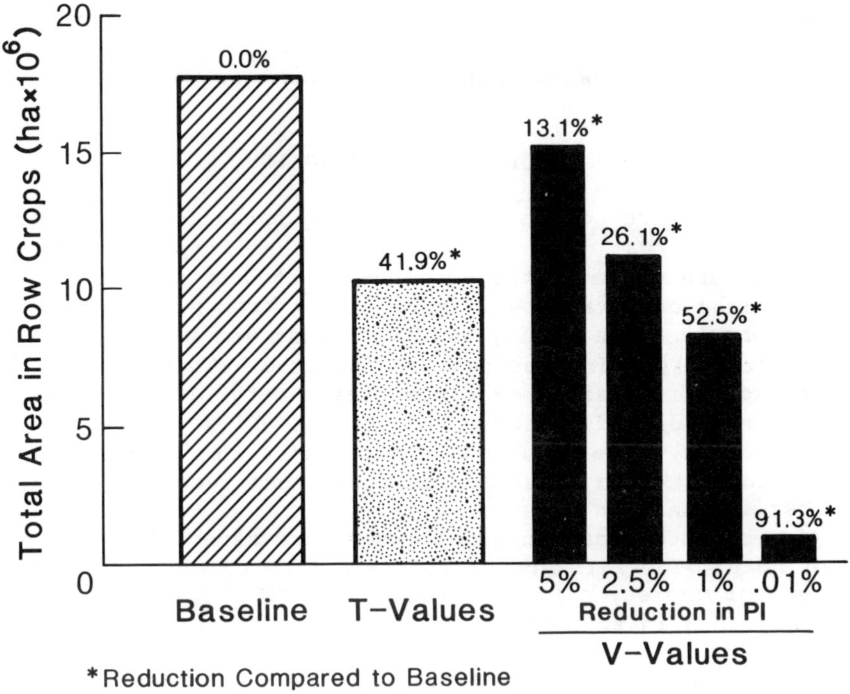

FIGURE 3 Total acres in row crops for all MLRAs under each scenario.

than the customary T value. Although use of V values does not eliminate the need for shifts in land use, when 100-year productivity losses are set at 5.0 and 2.5 percent these shifts are far fewer than implied by T values alone (see Figure 3). Not only are fewer acres likely to be targeted for land use changes, but the particular acres chosen are more likely to exhibit specific soil characteristics damaging to long-term productivity.

To simplify the analysis, a complete shift from row crops to forage has been assumed. Less extreme changes in rotation can and should be encouraged, based on local economic and soil characteristics. When the conservative assumptions used in this study are modified, substantial improvements in soil productivity may result without major disruptions in land use, provided policies are properly targeted. Finally, it must be reiterated that corn and soybean production in 1982 nearly broke records for those crops, with many marginal acres in production. Use of 1982 as a baseline therefore may overstate the needed reductions in this acreage in other years.

IMPLICATIONS FOR POLICY

Use of more accurate targeting criteria for soil and water conservation policy can reduce onsite productivity losses and minimize the acreage affected by more restrictive land use practices. Acres taken from production can and should be targeted, and those that are most vulnerable to erosion can increasingly be isolated. This study provides preliminary evidence that a targeting criterion can be developed, based on the recent NRI and Soils-5 data bases. Clearly, many difficulties and questions remain, although these appear to be less technical than institutional in nature.

Two of the institutional issues are especially worthy of note. The first concerns the choice of an appropriate rate of depletion of soil resources. The results of Scenario 3 in this paper clearly demonstrate the importance of this judgment and the magnitude of its effect on land use policy. Analysts are likely to differ over this rate, and no simple solution to the issue is possible (*see* Lind et al., 1982; Page, 1977). Nonetheless, current actions reveal an implicit rate that may well reflect existing preferences, as expressed by the 1982 baseline data reported above. Any policy applied with respect to targeting will have productivity implications over time and will reveal a similar implicit rate of depletion. It would be best, however, to make these judgments explicit. The vulnerability criterion developed in Scenario 3 does this. The choice of a "steady-state" rate, for example, appears to have major implications for future row-crop production in the Midwest, as is clearly revealed by this analysis.

The second institutional issue worth noting concerns the impact of targeting on existing Soil and Water Conservation Districts and the wide range of other institutions developed since the 1930s to deal with related land use issues. Some have argued that targeting would make these institutions less important; one consequence of this has been the arousal of opposition to the targeting concept. Yet, targeting of soil and water conservation policy does not diminish the important role of these institutions. Rather, it changes their role to one in which programs are more accurately directed and specifically fashioned to suit local needs. This implies, if anything, a broadened set of responsibilities for existing institutions, with ever greater emphasis on local autonomy over land use decisions based on improved technical information.

Finally, the preliminary nature of these findings must again be emphasized, along with the need for continued improvements in technical methods to identify onsite productivity losses due to soil erosion. These losses are, of course, only one aspect of a larger problem that includes important offsite damages (see Christensen, this volume; Crosson and Stout, 1983). It seems, however, that important beginnings can be made by estimating onsite damages, with further and more difficult estimates of offsite damages to follow. As technical capabilities increase, a similar commitment to institutional innovations can result, leading to reduction in productivity losses arising from poor management of America's great inherited wealth: her soil resources.

REFERENCES

Crosson, P. R., with A. T. Stout. 1983. Productivity Effects of Cropland Erosion in the United States. Baltimore, Md.: Johns Hopkins University Press for Resources for the Future.

Ervin, D. E., W. D. Heffernan, and G. P. Green. 1984. Cross compliance for erosion control: Anticipating efficiency and distributive impacts. Am. J. Ag. Econ. 66:273-278.

Larson, W. E., F. J. Pierce, and R. H. Dowdy. 1983. The threat of soil erosion to long-term crop production. Science 219:458-465.

Lind, R. C., K. J. Arrow, G. R. Corey, P. Dasgupta, A. K. Sen, T. Stauffer, J. E. Stiglitz, J. A. Stockfisch, and R. Wilson. 1982. Discounting for Time and Risk in Energy Policy. Baltimore, Md.: Johns Hopkins University Press for Resources for the Future.

Page T. 1977. Conservation and Economic Efficiency: An Approach to Materials Policy. Baltimore, Md.: Johns Hopkins University Press for Resources for the Future.

Pierce, F. J., W. E. Larson, R. H. Dowdy, and W. A. P. Graham. 1983. Productivity of soils: Assessing long-term changes due to erosion. J. Soil Water Conserv. 38:39-44.

Pierce, F. J., R. H. Dowdy, W. E. Larson, and W. A. P. Graham. 1984. Soil productivity in the Corn Belt: An assessment of erosion's long-term effects. J. Soil Water Conserv. 39:131-136.

USDA (U.S. Department of Agriculture). 1983. Soils-5 Data Base. Ames, Iowa: Soil Conservation Service.

Walker, D. J. 1982. A damage function to evaluate erosion control economics. Am. J. Ag. Econ. 64:690-698.

APPENDIX

Definitions

Universal Soil Loss Equation (USLE)

The USLE is described in detail by Wischmeier and Smith (1978). The equation was developed from more than 10,000 plot years of basic runoff and soil erosion data measured at 49 research locations in the United States as well as data obtained from rainfall simulator studies. The equation takes the form:

$$A = RKLSCP, \qquad (1)$$

where A is the erosion rate in tons/acre/year. R is the rainfall and runoff factor and is the number of rainfall erosion index units, plus a factor for runoff from snow-melt or applied water where such runoff is significant. K is a soil erodibility factor, which expresses the rate of soil removed per erosion index unit for a slope of specified geometry. L is the slope-length factor, representing the ratio of soil erosion from the field slope length to that from a standard length (72 feet) under identical conditions. S is the slope-steepness factor and is the ratio of soil erosion from the field slope gradient to that from a 9 percent slope under identical conditions. C is the cover and management factor and is the ratio of soil erosion from an area with a specified cover and management factor to that from an identical area in tilled, continuous fallow. P is the support practice factor and is the ratio of soil erosion with a support practice like contouring, strip-cropping, or terracing to that with straight-row farming up and down the slope.

Erosion Potential (EP)

The right side of the USLE can be separated into two parts: the factors that are controlled by nature or are affected by humans only at great costs (RKLS) and the factors governed essentially by management (CP). RKLS is therefore the inherent potential for erosion at a given location. The actual erosion rate (A) depends on the values taken by C and P.

Soil Loss Tolerance Limit (T)

This term is defined by Wischmeier and Smith (1978) as the maximum level of soil erosion that will permit a high level of crop productivity to be sustained economically and indefinitely. When substituted for A in the USLE, it allows an estimate of the maximum CP value necessary to keep erosion rates below the tolerance level, once RKLS is considered as a constant for a given location. T values range from 2.2 to 5.0 tons/acre/year and are based on experience and observations established through six regional workshops in 1961 and 1962 (Wischmeier and Smith, 1978).

Productivity Index (PI)

The PI used in this study is a modification by Pierce et al. (1983) of a model developed by Kiniry et al. (1983). The model indexes the soil according to its suitability as an environment for root growth. It was modified to include some additional concepts and to use data available in the Soils-5 data base (USDA, 1983). The modified model is:

$$PI = \sum_{i=i}^{r} (A_i \cdot C_i \cdot D_i \cdot WF), \tag{2}$$

where A_i is sufficiency of available water capacity, C_i is sufficiency of bulk density (adjusted for permeability), D_i is sufficiency of pH, WF is a weighting factor and r is the number of horizons in the depth of rooting. The model assumes that nutrients are nonlimiting to plant growth and that other factors are constant.

Soil Vulnerability (V)

The relative vulnerability of a soil to long-term erosion losses can be assessed by the slope of a PI-soil removal curve (Pierce et al., 1984), estimated as:

$$V = \Delta PI/\Delta d, \tag{3}$$

where ΔPI is the percentage variation (%) in PI and Δd is the change in depth (cm) due to soil erosion. Pierce et al. (1983) used a constant arbitrary Δd of -50 cm, which was also adopted in this paper. Although a few values were equal to or greater than zero, the present study reports V as an absolute number.

Simulation Methodology

Scenario 1: Baseline

The 1982 NRI furnished the erosion rate (A_1), the C and P values, and the coded soil unit corresponding to each sampling location within a MLRA. The coded soil unit was matched with the proper unit stored in Soils-5, which allowed the calculation of PI by Equation 2 using $A = A_1$.

The variation in PI over time (ΔPI, %) was estimated by:

$$\Delta PI = (PI_0 - PI_1)\ 100/PI_0, \qquad (4)$$

where PI_0 is the productivity index at time zero (here, 1982) and PI_1 is the productivity index after the removal of soil corresponding to 100 years of erosion at the present rate (A_1). The C and P values were multiplied together and reported as CP values. All results were weight-averaged by acreage for each MLRA.

Scenario 2: Use of Soil Loss Tolerance Limit Value

Using the same basic data as in Scenario 1, each sampling location had its erosion potential (EP) calculated as $EP = A_1/CP$ and its A_1 value compared to its T value furnished by the 1982 NRI. All locations in which A exceeded T were assumed to be taken out of row crops and the acreage was summed and then deducted from the total row-crop acreage in Scenario 1. The acreage taken out of row crops was then assigned a CP value of 0.01, corresponding to an established, well-managed forage field with about 80 percent ground cover (Wischmeier and Smith, 1978). For these locations, a new erosion rate (A_2) was calculated as $A_2 = EP \cdot CP$, using CP = 0.01. The PI values for acreage remaining in row crops as well as that shifted to forage was then calculated as in Scenario 1, and all results were weight-averaged by acreage for each MLRA, to yield productivity losses over a period of 100 years.

Scenario 3: Use of Soil Vulnerability Values

Vulnerability values were calculated according to methods developed by Pierce et al. (1984) for each soil sampling location planted to corn and soybeans in each MLRA. These V values and the four different degrees of productivity loss were used to calculate acreage that would be taken out of row crops and put into forage. Changes in PI (ΔPI) of 5.0, 2.5, 1.0, or 0.01 percent over 100 years were used to calculate Δd using Equation 3. This allowed the calculation of the erosion rate (A_t) for a tolerable reduction in PI by:

$$A_t = (\Delta d \cdot PI \cdot W)/(V \cdot t), \quad (5)$$

where W is the weight of a soil layer 1 acre in area and 1-inch thick, determined using the bulk density value for the local soil series reported in Soils-5, and where t is time (here, 100 years). Equation 5 is a modification of Equation 3 in Pierce et al. (1984).

The A_t values were then compared with local A_1 values. Where A_1 exceeded A_t, the area was taken out of row crops and assigned a C value of 0.01, as in Scenario 2. For these locations a new erosion rate (A_3) was determined as $A_3 = E_p \cdot CP$, with EP calculated as in Scenario 2 and CP equal to 0.01. The acreage taken out of row crops was again deducted from the total row-crop acreage in Scenario 1. All results were weight-averaged by acreage using the new values where appropriate.

REFERENCES

Kiniry, L. N., C. L. Scrivner, and M. E. Keener. 1983. A soil productivity index based upon predicted water depletion and root growth. Res. Bull. 1051. Columbia: Missouri Agricultural Experimental Station.

Pierce, F. J., W. E. Larson, R. H. Dowdy, and W. A. P. Graham. 1983. Productivity of soils: Assessing long-term changes due to erosion. J. Soil Water Conserv. 38:39-44.

Pierce, F. J., R. H. Dowdy, W. E. Larson, and W. A. P. Graham. 1984. Soil productivity in the Corn Belt: An assessment of erosion's long-term effects. J. Soil Water Conserv. 39:131-136.
USDA (U.S. Department of Agriculture). 1983. Soils-5 Data Base. Ames, Iowa: Soil Conservation Service.
Wischmeier, W. H., and D. D. Smith. 1978. Predicting rainfall erosion losses: A guide to conservation planning. Agriculture Handbook 537. Washington, D.C.: U.S. Department of Agriculture.

Discussion

John A. Miranowski

Runge, Larson, and Roloff have presented a framework for policy analysis that provides an excellent beginning in a new area of conservation policy research--using physical measures of soil productivity loss as the basis for soil conservation policy decisions. Simply targeting erosion control to the most erodible acres may not be the most efficient approach to the soil erosion problem. Gross soil loss may not be an accurate reflection of the potential productivity foregone. Using a measure of productivity loss, such as produced by the PI (productivity index) model (employed in this analysis) or the EPIC (the Erosion-Productivity Impact Calculator) model, provides a more logical basis for targeting erosion control.

The vulnerability index or measure discussed by Runge, Larson, and Roloff does allow for social judgment regarding the level of productivity loss that the public is willing to tolerate on croplands. But to some extent the vulnerability index criterion ignores the so-called "hard-core" economic information that should enter into this judgment.

The vulnerability index does not provide a complete accounting of the added benefits and costs that may be involved. First, the decision to retire cropland acres from row-crop production should be based on social benefit-cost calculus. If the added benefits outweigh the added costs of retirement, then the policy can be justified in an economic sense. If not, society's welfare is reduced. Without a more explicit accounting of the costs incurred, it is difficult to ascertain how society will determine the allowable rate of productivity decline, or as Runge and coauthors state, "it is our view that these judgments must reflect both private and social values concerning the appropriate rate of depletion of soil resources."

Second, retiring acres that are eroding at rates that lead to productivity losses greater than 1.0, 2.5, or 5.0 percent will affect a significant portion of cropland, as their Table 2 shows. Such large cropland retirements will have significant price effects, which in turn will have impacts on the mix of crops and tillage practices used on less erodible cropland. These adjustments may create related erosion problems but of lesser magnitude on the remaining cropland acres.

In cases where highly productive soils are found on highly erodible lands, it may nevertheless be appropriate from an economic perspective to continue farming these soils but to employ more intensive conservation practices. Retirement is not necessarily the most efficient alternative. Some recent work in Iowa (Miranowski and Hammes, 1984) considered land purchasers' willingness to pay for topsoil depth and erodibility (measured by the RKLS). Holding topsoil depth constant, the value of farmland decreased as erodibility increased. Landowners make investment decisions with respect to land purchases and conservation investments that reflect these tradeoffs between productivity and erosion control cost. It may prove costly to retire highly vulnerable cropland that is also highly productive. Additionally, some soils, precisely because they are highly productive and too costly to save, may be mined during periods of high commodity prices. These factors should be considered in any social decision to protect productivity of specific cropland through retirement programs.

The vulnerability measure proposed by Runge, Larson, and Roloff also appears to ignore technological change and potential soil genesis. These omissions would tend to overstate vulnerability and the need for policy intervention. The vulnerability measure may also ignore nonlinearities, because it is looking at the marginal increment at a particular point in time. Unless targeting is continually readjusted over time, program managers using the vulnerability measure may initially fail to retire those croplands with a nonlinear productivity decline relationship, thus underestimating the need for policy intervention.

It is also important to remember that there are two soil conservation goals: maintaining productivity, which is very important, and avoiding the offsite impacts of erosion, which may be of even greater benefit to society (Clark et al., 1985). Measures of these offsite impacts should be integrated into the analysis as well.

Finally, the economics profession needs to be challenged to focus greater attention on the economics of the soil erosion problem. Economists have made a major contribution to measuring productivity losses because they come from a tradition that tends to look for common factors to explain systematic behavior, that has been willing to aggregate individual decisions and draw broader generalizations, and that has emphasized model development and simulation. Although modeling and interpreting productivity impacts is an important first step, sight cannot be lost of the economic issues and the need to measure the dollar costs and benefits that are involved. In other words, we cannot afford to become too enamored with the indices and physical productivity measures that are being calculated or developed. Rather, creativity is needed in translating physical measures into economic measures, i.e., the net economic benefits of erosion control programs and policies. It is crucial that these productivity measures are now taken through the final step of the analysis to determine which soils should be targeted, retired, or saved because the social benefits of the policy action exceed the social costs.

REFERENCES

Clark, E. H., J. A. Haverkamp, and W. Chapman. Eroding Soils: The Off-Farm Impacts. Washington, D.C.: The Conservation Foundation.

Miranowski, J. A., and B. D. Hammes. 1984. Implicit prices for soil characteristics for farmland in Iowa. Amer. J. Agr. Econ. 66:745-749.

11

Potential Uses of the NRI in State and Local Decision Making

Chris J. Johannsen

Many people have been impressed by the beautiful landscapes of Missouri and they soon find that there are many interesting contrasts. Missouri is indeed a state where the north meets the south and east meets the west in terms of geology, soils, crops, climate, and people. Looking at these different landscapes, it is sometimes difficult to determine if anything is wrong. Yet, when something is very wrong, even the untrained eye can observe it. Missouri has a very serious soil erosion problem.

Missouri ranks second in the country behind Tennessee in rates of erosion from cropland acres. Its erosion rate is over 10 tons of soil loss/cropland acre, which amounts to nearly 160 tons of soil loss each year (USDA, 1981). During 1982, with the heavy spring rains, the estimated soil loss was nearly 22 tons/acre (Johannsen, 1982). That phenomenon was repeated this last spring--indeed, the estimate increased to over 25 tons/acre.

The northern part of Missouri has experienced a loss of cattle production operations resulting in many pasturelands being plowed and put to row crops. In Atchison County alone, satellite images recently documented that 70 percent of the county area is now in row crops. This is in contrast to 5 years ago, when less than half the county was in such crops. During the past 5 years, many counties in northern Missouri have lost from 35 to 50 percent of their pastured areas to row-crop production.

The Deep Loess Hills in northwest Missouri have experienced erosion rates as high as 50 tons/acre/year. Many road ditches have been completely silted in. It is interesting to note that the county highway engineers were some of the first to complain about the ditches being filled; they did not have enough money budgeted to remove the materials.

In the west central part of the state, the productive soils are developed on acid shale materials with loess or windblown topsoil. This area, called the Cherokee Prairie, has many severe erosion problems, but it receives little publicity because its erosion rates are not as dramatically high as those of the northwest portion of the state. This nearly level to gently sloping area is intensively farmed and, with low water intake due to slightly heavy textured surface soil, erosion rates are normally high. The loss of topsoil and exposure of some acid subsoils will be a very serious problem to the agricultural economy of this area with continued erosion rates.

Missouri's Ozark region has bottomland areas that are usually row-cropped. Many of the upland areas have very shallow soil with little topsoil. Over 2 million acres of forest and brushland have been aerial sprayed with herbicides or removed by bulldozer and then seeded to grass. Some of these areas have been intensively grazed, resulting in severe erosion. Any erosion is a major problem since the topsoils are thin and subsoil materials are usually very infertile.

The Bootheel area of the state contains soils that are formed in alluvium, either on floodplain or terraced positions. The soils are deep and vary in texture from clay to loamy sands. The problem receiving increasingly more attention in this area is wind erosion.

SEEKING A SOLUTION

In 1981, the Soil Conservation Service (SCS) of the U.S. Department of Agriculture (USDA) held a meeting with state and federal resource agencies, farm organizations, and the University of Missouri to assess the resource information needs of these groups. As a result, Missouri expanded the data collection format of the proposed statewide National Resources Inventory (NRI) to include more detailed wildlife habitat and timber resource data. Additionally, at the request of the participants, the data were put in a geographically referenced data base to display queries in graphic, map, and tabular formats. The Geographic Resources Center (GRC) at the University of Missouri-Columbia was contracted to develop the NRI geographic data base.

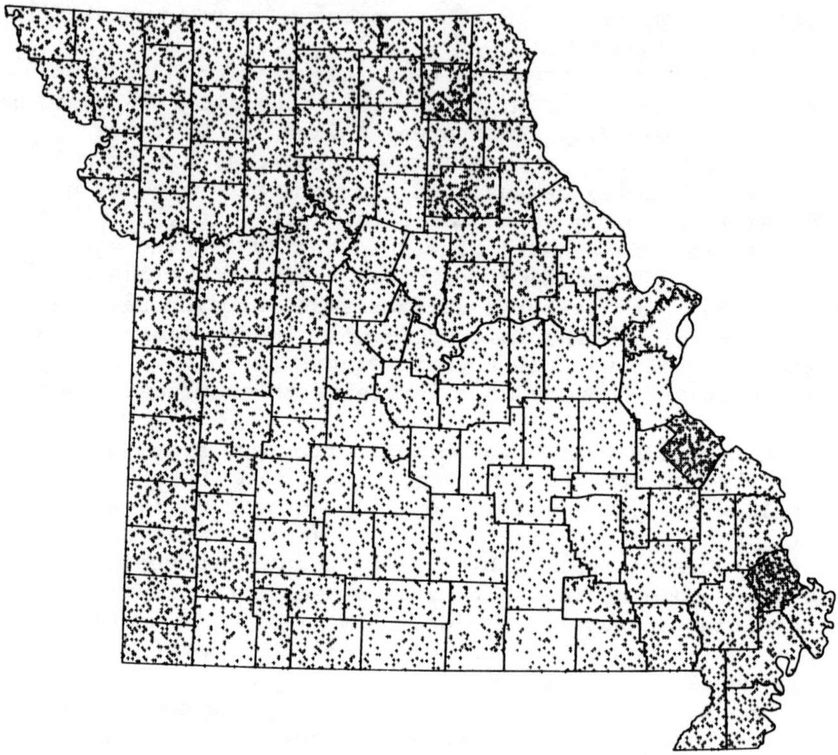

FIGURE 1 Primary sampling unit (PSU) locations in Missouri.

The samples selected for the 1982 NRI in Missouri totaled over 13,000 primary sampling units (PSUs), shown in Figure 1. These represented a sample of about 4 percent and translated to 150 PSUs for an average Missouri county. Since three sample locations were selected statistically within each PSU, data were collected from over 39,000 locations by SCS soil scientists and district conservationists, providing the most detailed data ever collected on Missouri's soil and water resources (Johannsen, 1984).

The development of the PSU data base was divided into four steps: data entry and verification, digitizing PSU locations, digitizing ancillary data, and implementation of a retrieval and display system (Johannsen et al., 1984). Data entry and verification will be a continuing process until about 15,000 PSU points are collected and entered into the data base. To date, approximately 13,000 PSUs have been entered into the system. Additional samples are being obtained to provide county-reliable data for more than 25 of Missouri's 114 counties.

The field data worksheets were entered with a specially designed data entry program through the use of form-drive interactive mechanisms. Every form displayed on the operator's terminal was constructed to look just like its corresponding section on the original worksheet. Twenty-one screen forms are required to cover an entire PSU worksheet. When a form has been entered, the next one is prompted automatically. Filled forms are easily recalled for editing or correction.

The individual PSUs were located on Geological Survey 7.5-minute quadrangle sheets and digitized to describe accurately their locations in Universal Transverse Mercator (UTM) coordinates for use by the data retrieval and display system. In addition, seven statewide maps were digitized for inclusion in the data base; they serve as supplemental geocodes for improving the analytical and reporting functions of the systems. The digitized maps include general soils, county boundaries, forest cover, zoographic regions, fish-fauna regions, Major Land Resource Areas, and hydrologic units. The maps were digitized manually. After capturing the line segments, the segments were cycled to produce polygons that can be displayed with the locations of selected PSUs on a raster or vector graphics device.

A relational data base management system integrates the PSU field data, PSU locations, supplemental attribute geocodes, and supplemental attribute boundary files (Johannsen et al., 1984). The PSU field data file and supplemental attribute file are used to create a retrieval file that is searched to answer all queries. The PSU location file and supplemental attribute boundary file are primarily used to put retrieval results into graphic format on a display monitor or plotter. All search results can be reported in tabular or graphic formats.

USING THE PSU DATA BASE

The Soil Conservation Service (SCS) and the Office of Soil and Water Conservation Programs of the Missouri Department of Natural Resources are currently requesting specific illustrations and results from the Missouri PSU data base. In August 1984, Missouri voters approved a sales tax of 0.1 percent to provide state funds for a cost-sharing program as an incentive to landowners to use conservation practices. The funds will be administered through the Department of Natural Resources, with SCS providing technical assistance. Funds are to be used when USDA/Agricultural Stabilization and Conservation Service (ASCS) funds are not available for erosion control practices. Therefore, soil and water districts are cooperating with county ASCS committees in the program.

The funds will be distributed on the basis of a formula that provides a portion to each district, with the remainder allocated according to the percentage erodible cropland acreage of the state total. Figure 2 shows the distribution of cropland PSUs and the location of cropland PSUs with losses greater than 10 tons/acre/year. The latter was determined using the Universal Soil Loss Equation (USLE), which was calculated from the data collected in the field for each point location within a PSU. The data could be requested on the basis of soil loss tolerance limit (T value), twice the tolerance (2T), or any specific erosion level entered by the user.

The flexibility of the retrieval of the statewide data is shown in Figure 3, where the user wanted to see row-crop and small-grain PSU locations. The locations of a specific crop can also be retrieved. For example, the location of soybeans was requested by the State Extension Agronomy Specialist with responsibility for that crop. Warm season grasses (see Figure 3-C) are of increasing interest to wildlife specialists, while the distribution of forests (see Figure 3-D) that are being grazed are of interest to the resource professionals working with the timber industry.

Maps and illustrations are being provided to each SCS planning area such as those shown in Figure 4. Other agencies with different area boundaries, such as the Extension Service and ASCS, can request illustrations by providing an organizational map or listing of counties to the Geographic Resources Center. The data can also be presented in a table (such as the soil loss by land capability class in Table 1) or by pie chart (such as Figure 5, which shows the distribution of different crops in SCS Area 3). Additionally, bar charts like those shown in Figure 6 can be developed for comparing data between counties or any categories selected by the user.

County-reliable data have been collected for about 15 percent of Missouri's counties. One example is Monroe

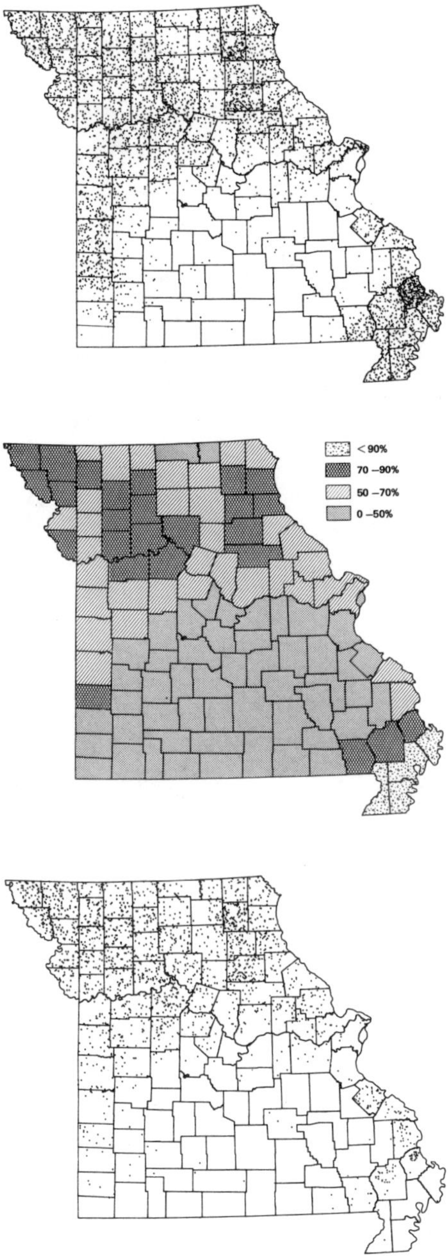

A. Cropland PSUs with county boundaries

B. Cropland distribution

C. Cropland with greater than 10 tons/acre/year soil loss

FIGURE 2 Cropland PSUs in Missouri with information presented in different formats.

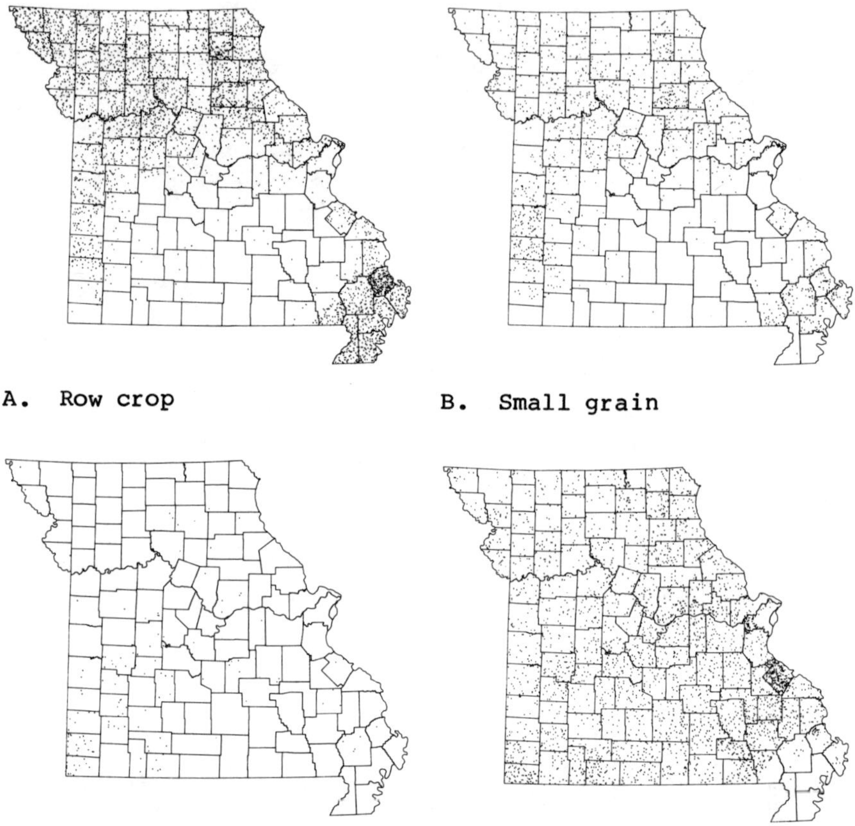

A. Row crop

B. Small grain

C. Forage with warm season grasses

D. Forest and grazed forest

FIGURE 3 Sample statewide data retrieved by PSU location, specifying land cover or land use categories.

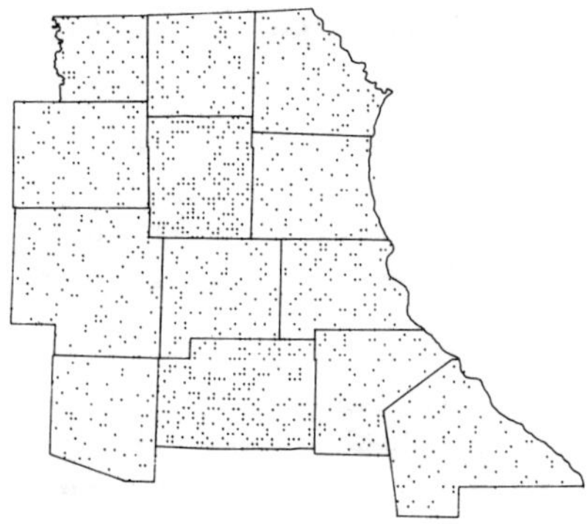

A. PSU locations

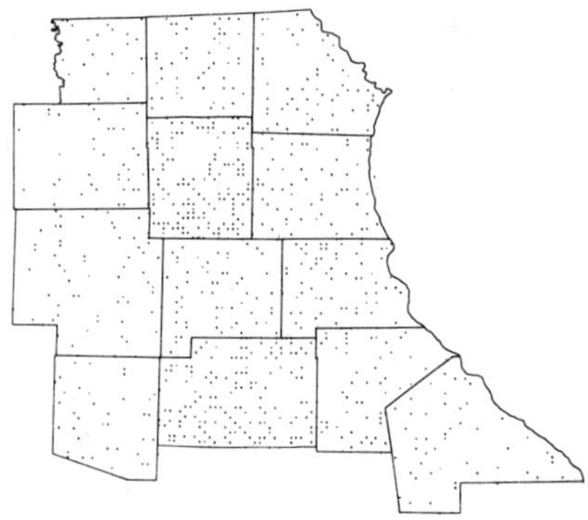

B. Cropland PSUs

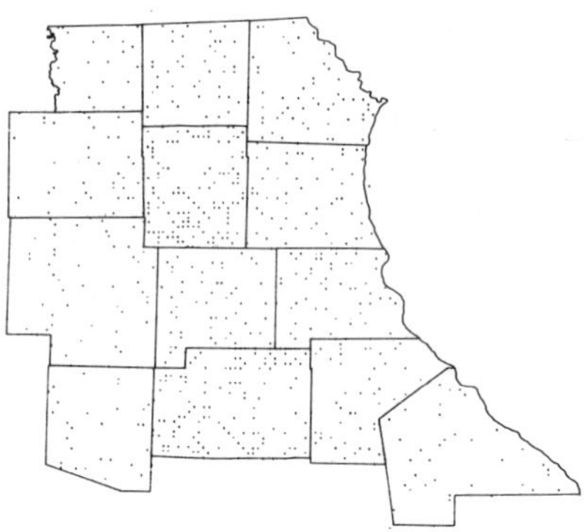

C. Row crop PSUs

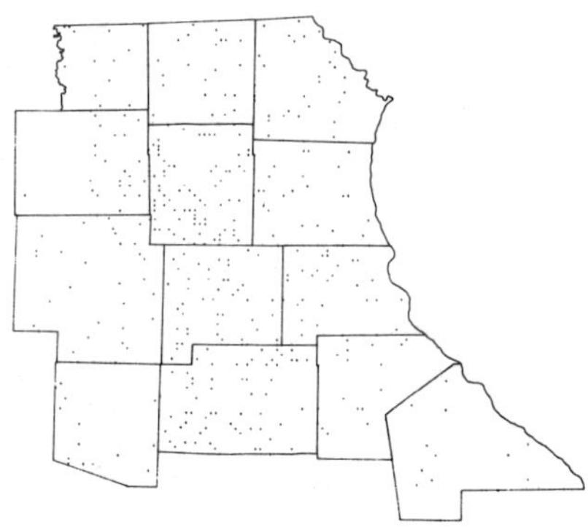

D. Row crop with greater than 10 tons/acre/year soil loss

FIGURE 4 Use of NRI data, SCS Area 3, northeast Missouri.

TABLE 1 Average Soil Loss and Land Use Categories by Soil Loss Tolerance (T) for Area 3 in Northeast Missouri

Land Use	Soil Loss Tolerance	Average Soil Loss[a] (Tons/Acre)	Acres (100)
Cropland	<T	2.42	6,317
	T to 2T	5.08	4,017
	>2T	18.79	11,869
	All	11.99	22,203
Grassland	<T	1.20	8,339
	T to 2T	5.39	1,648
	>2T	16.48	1,996
	All	3.99	11,983
Forest	<T	0.63	5,204
	T to 2T	4.76	583
	>2T	24.90	680
	All	3.06	6,467
Grazed forest	<T	1.28	1,016
	T to 2T	5.13	394
	>2T	25.93	562
	All	7.95	1,972
Nongrazed forest	<T	0.46	4,188
	T to 2T	3.95	189
	>2T	20.43	118
	All	1.04	4,495
Urban	All	0.00	45
Other	All	9.87	783

[a]Calculated by Universal Soil Loss Equation using data collected from PSU locations.

SOURCE: 1982 NRI.

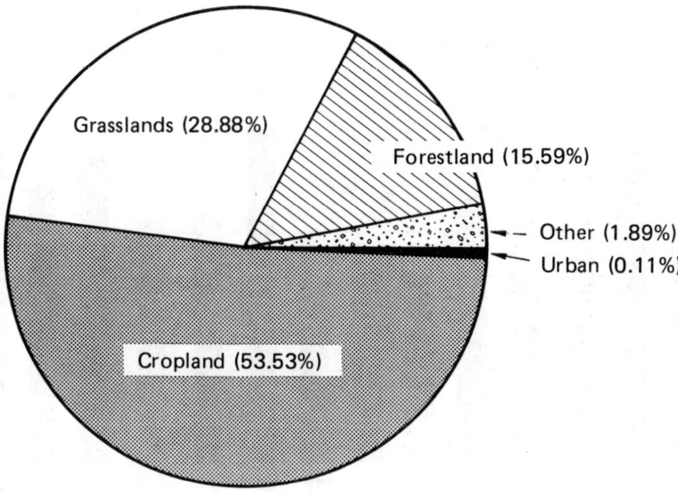

FIGURE 5 Distribution of cropland by a specific crop in Area 3.

County (see Figure 7), located in northeast Missouri. The data can be retrieved by watershed boundaries or by cropland classes and specific conservation treatment. The query for information is additive so that the user can request multiple conditions from the data. Individual county displays will only be made for counties with county-reliable sample points so that information provided is not misused. Table 2 illustrates specific data for Monroe County, Missouri, showing the amount of row-crop acreage on different land capability classes. This county may want to target its funds to assist either the most productive lands or the areas that should be removed from cropland production.

All the soil and water conservation districts are in the process of developing 5-year plans. Information from all available sources and the NRI data will be used to establish priorities for starting conservation. All districts have funds available to help implement their plans--a resource which increases emphasis on developing sound plans. The district conservationist and local extension specialist, a district supervisor by virtue of the position, are working together to write the plans with assistance from their elected supervisors.

There are many possibilities for display and use of the NRI data in Missouri. Many uses will only be known

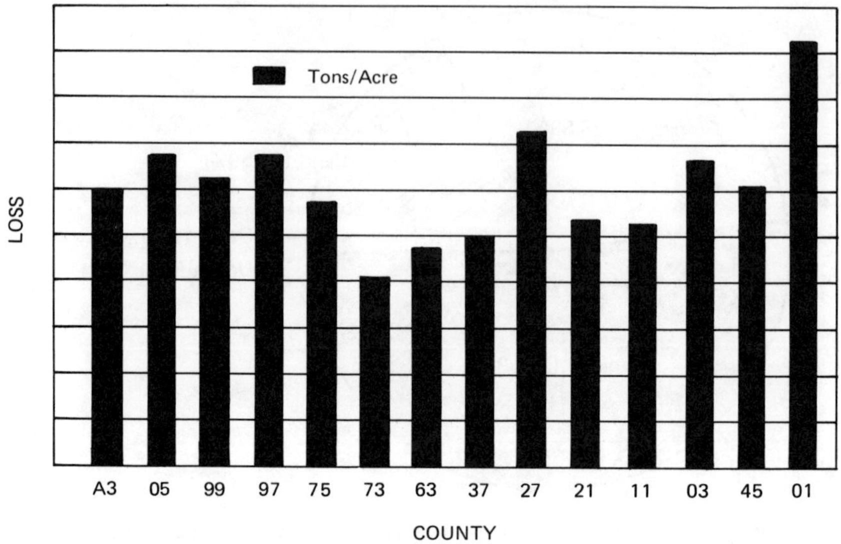

FIGURE 6 Average soil loss for row crops for selected counties in northeast Missouri. Key: Average of all 13 counties, A3; Knox, 05; Shelby, 99; Scotland, 97; Schuyler, 75; Randolph, 73; Ralls, 63; Monroe, 37; Marion, 27; Macon, 21; Lewis, 11; Clark, 03; Pike, 45; Adair, 01.

as people become familiar with the data base. A recent request for the locations of irrigated lands by one of the state agencies, for example, brought an additional appreciated response when it was learned that information on the source of the water could also be provided.

SUMMARY AND CONCLUSIONS

The data base management system established for analyzing Missouri's NRI data serves as a model for query, display, and application of these data in other states. Further data collections of this magnitude will likely consider a geographic information system approach in planning data collection and use of results.

Missouri is in the initial stages of using the results from this system. Many uses of the products have not yet been envisioned. The flexibility of retrieving the results in formats requested by the user will lead to

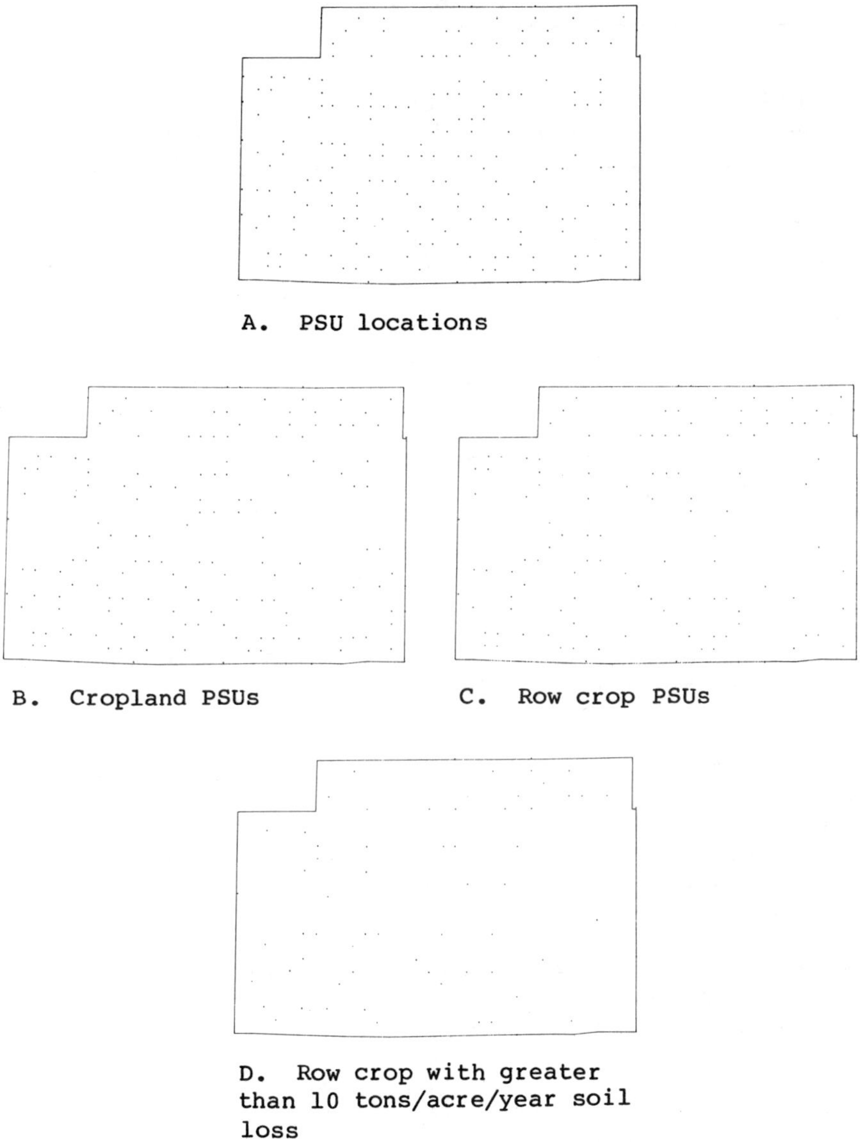

FIGURE 7 Use of NRI data for Monroe County (northeast Missouri) with county-reliable samples.

TABLE 2 Average Soil Loss and Acreage by Land Capability Class for Monroe County, Missouri

Land Capability Class	Subclass	PSU Points	Average Soil Loss (Tons/Acre)	Acres (100)
II	e	124	9.56	731
	w	86	5.71	519
III	e	117	12.61	731
	w	11	5.37	65
IV	e	28	13.86	172
VI	e	2	31.05	16
Total			9.95	2,234

SOURCE: 1982 NRI.

many additional cooperative efforts among resource agencies that are trying to stop the deterioration of Missouri's soil resource and its influence on vegetation, water, and wildlife resources.

REFERENCES

Johannsen, C. J. 1982. Soil erosion of Missouri: An inventory. Pp. 13-15 in Soil and Water Conservation: The Principle and the Practice. Special Report 290. Columbia: Agricultural Experiment Station, University of Missouri.

Johannsen, C. J. 1984. Missouri's National Resources Inventory. University of Missouri-Columbia Agronomy Technical Report 2(11):5-7.

Johannsen, C. J., J. M. Pan, T. W. Barney, and G. T. Koeln. 1984. A data base management system for Missouri's National Resource Inventory. Pecora IX Symposium, Sioux Falls, S. Dak., Oct. 12-15.

USDA (U.S. Department of Agriculture). 1981. Soil and Water and Related Resources in the United States, 1980 RCA Appraisal, Part I. Washington, D.C.: U.S. Department of Agriculture.

Discussion

Max Schnepf

Some states now realize the wealth of information they possess from the 1982 National Resources Inventory (NRI) data. Missouri is a good example, as Johannsen's paper points out. Other states and local governments have yet to recognize how flush they are in terms of the value of the NRI data.

OVERALL OBSERVATIONS

The NRI is a national assessment tool, first and foremost, and while the 1982 NRI is accurate to the level of Major Land Resource Area (MLRA), there are real limitations to its use at the county level and, for some purposes, even at the state and national levels.

The 1977 NRI raised questions about the value of the data to individuals and agencies at the state and local levels. Many bought into the 1977 NRI with their time and effort and apparently wanted more out of the assessment than they felt they received. As a result, data accurate to the county level became an objective of the 1982 NRI. When budget considerations forced abandonment of that objective, the compromise was data reliable to the MLRA level.

A number of states--Missouri, Louisiana, and Kansas, for example--opted to spend additional funds to achieve county-level accuracy in all or selected counties, and about 200 counties now have reached this goal. Several hundred more are in the process of collecting data to achieve this level of accuracy.

The fact that MLRA boundaries do not coincide with county boundaries poses problems for use of the data at the county level. In some cases, as Johannsen points out regarding Missouri, county-level information can be interpolated from MLRA data, but the accuracy of that interpolation is often questionable.

It will be difficult to avoid some misuse of NRI data, particularly at local levels, where the view surely will be "any information is better than no information." Comparison of 1982 NRI data with 1977 NRI data is possible in many cases at the state and national levels because the same PSUs (primary sampling units) sampled in 1977 were sampled again in 1982. In other cases, the data are not comparable because the methodology changed. An example is the case of urban and built-up uses of land. Because people want time-series data, the possibility exists in these latter cases of comparing apples to oranges.

The fact that the NRI excludes federal lands is a real shortcoming for state and local users in areas where there are extensive federal land holdings. Another problem is that the accuracy of the NRI data varies. For example, data on riparian land and wildlife habitat diversity are so limited that they are not accurate at the state level and may not even be so at the national level. Lastly, NRI data are of little use in dealing with relatively site-specific situations--for example, water quality problems emanating from livestock or poultry operations.

STATE AND LOCAL USES OF THE NRI

How has the NRI been used thus far by state and local decision makers? Beyond the experience in Missouri, information received from a number of directors of state soil and water conservation agencies and Soil Conservation Service (SCS) state conservationists paints a varied picture, for several reasons. First, the preliminary nature of the NRI data in some states has generated a cautious attitude among some administrators toward releasing the information and encouraging its use. Second, some states have created more elaborate data bases of their own, which are being used in conjunction with or in lieu of the NRI.

Missouri is unquestionably a leader in making use of the NRI. But there are other good examples of the data's use by state and local decision makers. Louisiana has perhaps been the most ambitious in developing and using NRI data. From the outset of the 1982 effort, the state pursued a much more elaborate sampling scheme that ensured data accurate to the county or parish level. A number of state and federal agencies in Louisiana are now using that data, as are some consultants and scientists at Louisiana State University (LSU). For example, the Louisiana Division of Water Pollution Control has used the information to identify the general location of soils with high erosion rates. The agency uses this data in its nonpoint source program to locate potential water quality problem areas resulting from soil erosion, and it recently expanded this effort to locate cropland and forested areas where erosion is serious and has the potential to pollute surface waters. The agency also uses the NRI data to locate areas with different crops growing on lands with serious erosion. This will help officials identify what water pollution problems might result from the use of fertilizers and pesticides on these crops.

The Louisiana Soil and Water Conservation Committee used the NRI data to help conservation districts develop workload analyses for staffing and budgetary needs. The NRI data are also used by the Department of Transportation and Development in its water quality management basin reports, by the Water Resources Studies Commission to identify potentially serious groundwater quantity and quality problems, by the Louisiana Department of Agriculture in selected public information programs, by the Corps of Engineers to identify land cover/use and soil erosion in watersheds of reservoirs, and by the Agricultural Stabilization and Conservation Service to estimate soil erosion on marginal lands by parish and MLRA.

The Department of Civil Engineering at LSU is using the NRI information in a pilot project at its Remote Sensing Laboratory to verify Landsat data. The LSU Department of Geography and Anthropology is using NRI data to develop a proximal mapping technique to show general locations of selected land use/cover and conditions of soil resources in the state. And LSU's Department of Agricultural Economics uses the NRI data to identify and analyze cropping patterns and yields by soil type in project areas.

In other states, use of the NRI data appears more limited:

- Indiana is completing a description of soil erosion problems in the state by the Governor's Soil Resources Study Commission.
- Alabama's Soil and Water Conservation Committee is developing a long-range soil and water conservation program for the state. The committee is seeking legislative support for an $8-million annual state appropriation for 20 years to fund this program.
- Nebraska's Natural Resource Commission is using the NRI in much the same fashion as Indiana and Alabama--to determine the magnitude of soil and water conservation problems in the state. The agency is also attempting to reorganize the data on the basis of its natural resource

districts, which follow hydrologic boundaries. It intends to allocate cost-sharing monies to the districts using the NRI data.

- Georgia is producing an information piece that uses NRI data to characterize the state's soil and water problems.
- Wisconsin is evaluating progress in soil erosion control and identifying priority areas for erosion control implementation projects.
- Illinois agencies are using the data as a basis for writing county and state "T by 2000" plans, which include treatment, personnel, financial, and extension needs.
- Kentucky is working on long-range soil and water conservation plans at the state level and in about half of the state's 121 conservation districts. The data were also used to establish parameters for development of a statewide soil erosion assessment computer model. And state officials used NRI information when seeking the governor's support for agricultural land protection initiatives.
- Kansas is developing a state water plan using NRI data. The NRI information is also being used to allocate state cost-sharing monies, and Kansas State University researchers now have tables of preliminary data available for research purposes.

Among other users of NRI data at state and local levels are the Agricultural Stabilization and Conservation Service, the Tennessee Valley Authority, and a number of researchers, several of whom report on their work in this volume. Some private groups, including the Texas Goat Raisers Association, have also requested the data.

POTENTIAL USES OF NRI DATA

The number of uses of NRI data in land and water planning is nearly endless. It is surely safe to say that the potential far exceeds what has been done to date. For example, the fact that the NRI data tapes can be meshed with the Soils-5 data tapes creates a number of analytical possibilities. As Johannsen mentions in his paper, there are possibilities for digitizing the NRI data and incorporating that data into geographic information systems. And putting NRI data into microcomputers has enormous possibilities for planning and decision making at state and local levels.

Obviously, what is needed most if the NRI data are to be used effectively and efficiently by state and local interests is a good educational effort. Thought must be given to creative ways of spurring interest in the NRI data, and then potential uses need to be advertised widely among state and local officials.

To date, there has been limited effort along these lines. All state offices of SCS have money in their 1984-1985 budgets to produce publications on the NRI. Many are assembling a technical document as well as a more popular brochure. The agency also has an on-line computer query system linking its Washington, D.C., office with its state offices. The considerable analytical work going on as a result of this link will benefit the states. In addition, SCS now has designated representatives in its four technical centers who are working with the states on the use of NRI data.

Some attempts are also being made within states to make potential users aware of the NRI data and encourage use of the information. Minnesota SCS officials, for example, recently sponsored a day-long seminar for federal, state, and local agencies and groups to acquaint them with the NRI and its use. Twenty-one agencies and groups were invited; 17 participated. In Kansas, SCS officials have discussed potential uses of the NRI at two meetings of that state's Agricultural Council, a group of farm- and ranch-oriented agencies and interest groups that meets monthly. Kansas and other states have also orchestrated broader publicity efforts via newspapers and other public media.

CONCLUSION

Three points seem worth making with respect to the future use of NRI information by state and local decision makers.

First, SCS could do much to extend the value of the NRI by establishing a formal or informal work group at the national level to do some creative thinking about the potential uses, to make a laundry list of those possibilities perhaps, and to set some guidelines on the appropriateness and accuracy of the data for various uses.

Second, the question of the need for and value of a broader national natural resource data base must be dealt with. The NRI information is of limited value in states with extensive federal land holdings. As the governor of one western state reportedly lamented, "the NRI tells me a lot about the privately held land in my state, but it doesn't tell me a darn thing about my state." If people at state and local levels are encouraged to use NRI information, some in the West in particular are going to want and need a more extensive data base than the NRI provides. Moreover, if they do indeed want and need that information, they are not going to be greatly concerned about the turf battles that occur in Washington, D.C., whenever the subject of a national data base is raised.

Third, as mentioned, many people at state and local levels bought into the NRI assessments in 1977 and 1982. Some still feel they are not getting as much out of those assessments as they should. As a result, pressure is likely to continue for data accurate to the county level in future NRIs. If the conduct of those assessments is to continue to rely on people at the state and local levels, some consideration must be given to accommodating their interests. The alternative is to devise an assessment process based on remote sensing or other technology that precludes the need for this state and local assistance.